Attitude Research in Science Education

Classic and Contemporary Measurements

Attitude Research in Science Education

Classic and Contemporary Measurements

Edited by

Issa M. Saleh
Bahrain Teachers College
University of Bahrain

Myint Swe Khine
Bahrain Teachers College
University of Bahrain

INFORMATION AGE PUBLISHING, INC.
Charlotte, NC • www.infoagepub.com

Library of Congress Cataloging-in-Publication Data

Attitude research in science education classic and contemporary measurements
/ edited by Issa M. Saleh, Myint Swe Khine.
    p. cm.
 Includes bibliographical references.
 ISBN 978-1-61735-325-3 (hardcover) – ISBN 978-1-61735-324-6 (pbk.) –
ISBN 978-1-61735-326-0 (e-book)
1. Science–Study and teaching (Secondary) 2. College
students–Attitudes. 3. Attitude (Psychology)–Testing. I. Khine, Myint
Swe.
 Q181.A836 2011
 507.1–dc22

                                                              2010049812

CONTENTS

PART I

INSTRUMENTS AND MEASURING SCIENCE ATTITUDE

v

PART II

SCIENCE ATTITUDE AND SOCIO-SCIENTIFIC ISSUES

PART I

INSTRUMENTS AND MEASURING SCIENCE ATTITUDE

CHAPTER 1

ATTITUDE RESEARCH IN SCIENCE EDUCATION

Norman Reid
School of Education,
University of Dundee

HISTORY OF ATTITUDES

The word "attitude" has a very vague range of meanings in normal day-to-day language. We often use the word to interpret almost any kind of behavior that is regarded as unusual or unacceptable. Indeed, attitudes may underpin many aspects of behavior, and that is why the study of attitudes is so important. However, in the world of social psychology, it took decades for the concept of attitude to crystallize into any kind of agreed meaning.

Herbert Spencer (1862, quoted by Ajzen & Fishbein, 1980) offered one of the first descriptions of attitudes when he spoke of "arriving at correct judgments on disputed questions much depends on the attitude of mind we preserve while listening to, or taking part in, the controversy" (p. 13). Perhaps Spencer was unfortunate in implying that there are "correct" judgments, but his idea of judgment is important.

The problem is that attitudes cannot be seen or observed directly. They exist in the brain. For many decades in the early 20th century, the dominance of behaviorist thought in the world of psychology meant that it was

Attitude Research in Science Education, pages 3–44
Copyright © 2011 by Information Age Publishing
All rights of reproduction in any form reserved.

3

not thought acceptable to seek to measure what could only be deduced by inference. The outcome of this was that attitudes were not regarded as an acceptable area of enquiry.

When Thurstone published his paper entitled "Attitudes Can Be Measured" (Thurstone, 1929), this offered a new challenge to the dominance of behaviorism. His approach was ingenious but very time-consuming (Thurstone & Chave, 1929). Nonetheless, his work opened up the whole area of attitude measurement, and he was quickly followed by the work of Likert (1932). Both Thurstone and Likert used a paper-and-pencil survey approach. The Likert approach asked people to look at a series of statements and to express their personal measure of agreement or disagreement with these statements using, originally, a seven-point scale from "very strongly agree" to "very strongly disagree." This is often modified today to a five-point scale, from "strongly agree" to "strongly disagree," and this style of question is probably the most common form of question used today.

Arising from a quite separate area of research, the work of Osgood, Suci, and Tannenbaum (1957) opened up a new approach that has proved extremely powerful. Osgood's team were really exploring what they described as semantic space. This can be loosely seen as the way we see ideas. Following a series of studies, they started to appreciate that the meaning of ideas could be crystallized down to three dimensions. They called these *evaluation, potency,* and *activity,* and these can be seen as the dimensions of *fast–slow, good–bad, powerful–powerless.* Their work is nicely summarized by Heise (1970). Osgood and his team appreciated that the *good–bad* dimensions related closely to attitudes, and this led to his approach being adopted widely as a tool in attitude measurement. The relative usefulness of these and other approaches will be discussed and illustrated later.

The whole area of attitude measurement received a considerable boost after World War II when considerable funding was made available in the U.S. for research. This led to a stream of research papers, and many of the findings from one major research team were brought together in the book *Communication and Persuasion* (Hovland et al., 1953). This text still offers key insights. However, much of the effort was focused on attitudes related to political stances or major social issues. This kind of attitude is very different from the kind of attitudes that are relevant in science education today.

THE CONCEPT OF ATTITUDES

We need to recognize that the concept of attitude has played an outstanding role throughout the history of social psychology (Ajzen & Fishbein, 1980). It is so central because attitudes influence patterns of behaviour. The problem is how we describe attitudes. Nearly 30 years ago, specifically in the field of science education, Johnstone and Reid (1981) observed that there were simply too many descriptions in the literature, resulting in a lack of clarity and also a lack of a common approach. The description that dominated was that of Allport (1935) who talked about "a mental and neural state of readiness to respond, organised through experience, exerting a directive and/or dynamic influence on behaviour" (p. 799). His definition has stood the test of time and has influenced many future thinkers and researchers.

Further refinements include those of Krech and Crutchfield (1948), Doob (1947), Katz and Sarnof (1954) and Osgood et al. (1957). Later, in 1958, Rhine referred to an attitude as a "concept with an evaluative dimension" (p. 364). The phrase *evaluative dimension* was important. It seemed to almost go back to Spencer's 1962 use of the word "judgment" and this "evaluative dimension" proposed by Rhine has assumed greater importance in later work. In some ways, this is what distinguishes an attitude from other latent constructs. A person may know, may have feelings, or may experience. However, it is possible that these may lead to evaluation and subsequent decisions. Eagly and Chaiken (1993, pp. 1–2) bring together many ideas when they state, "Attitude is a psychological tendency that is expressed by evaluating a particular entity with some degree of favour or disfavour." This description is now widely accepted.

There are several related words and this causes confusions. How do we consider opinions, beliefs, values? Are these the same as attitudes? In an ingenious analysis hidden away in a PhD thesis, Oraif (2007) offers an interesting analysis.

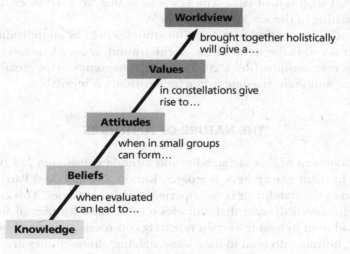

Figure 1.1 Hierarchy of ideas (after Oraif, 2007).

For our purposes here, the key thing is that Oraif clarified the distinctions between knowledge, belief, and attitude. She saw clearly that there was evaluation of knowledge involved in both beliefs and attitudes and that attitudes were formed from small groups of beliefs. She did not discuss opinions, but sometimes these might be seen as beliefs (very specific) or sometimes as attitudes (rather more general).

We can illustrate this with regard to physics. A school student may have studied some physics. This involves some knowledge of physics and some knowledge of the experiences of studying physics. On its own, neither is a belief or an attitude. However, in studying physics, the student may come to have negative feelings towards specific aspects of physics learning. Thus, the student may hold the belief that physics experiments often do not work, that the physics learned is unrelated to life, and that the teacher sets too much homework. These are beliefs but may well lead to an attitude towards physics that is negative. There is a negative evaluation of aspects of physics learning. In turn, such an attitude may lead to the rejection of further studies. Thus, we can see attitudes as being built up of a small set of beliefs. These beliefs may sometimes have some kind of internal consistency. Nonetheless, inconsistencies are possible.

Thus, in thinking of our school student in physics again, the person may well find aspects of the learning unattractive (a negative evaluation) while others are highly attractive (a positive evaluation). Thus, our student may view the laboratory learning positively, hold the teacher in warm regard while, at the same time, evaluate the physics taught very negatively. This has huge implications for all attitude measurement if it is to be useful, a point to which I shall return later. The key issue is that we need to explore the *detail* relating to the way attitudes are held.

Attitudes serve a vital purpose. The attitudes held by an individual help that person to make sense of the world around, sense of themselves and sense of relationships (Reid, 2003). While consistency helps greatly, life is complex, and there are times when inconsistency is inevitable.

THE NATURE OF ATTITUDES

The whole area of attitude stability and attitude consistency has been explored by many researchers. Roediger, Ruston, Capaldi, and Paris (1984) referred to the stability or relative permanence of attitudes. This contribution emphasized the fact that attitudes tend to have features of some stability and tend to lead to certain relatively consistent patterns of behavior. Indeed, attitudes do tend to have some stability, although they are open to development, modification, and change.

Cook and Selltiz (1964) noted that attitudes, on their own, do not control behavior, and this principle led to the development of the theory of reasoned action and the theory of planned behavior (Ajzen & Fishbein, 1980; Ajzen, 1985). I shall return to these theories later, for they raise important issues relating to attitude measurement.

Attitude consistency has been a major area of research. The early work of Heider (1944) laid the basis for this, and the ideas were picked up in the brilliant work of Festinger in the 1950s. Festinger (1957) showed in some amazing experiments that the extent of overall inconsistency between attitude and behavior offer an almost quantitative way to see when attitudes were likely to change and develop. His work has been replicated and was used in the 1970s in an early study on social attitudes arising from the study of chemistry. In this study, his ideas were applied and found to be well supported (Johnstone & Reid, 1981).

There is fairly general agreement (e.g., Bagozzi & Burnkrant, 1979; McGuire, 1985) that attitudes are stored in long-term memory and involve three components:

- What we know
- How we feel
- How we behave

This is more formally described in terms of attitudes having three components:

1. A knowledge about the object: the beliefs, ideas components (Cognitive).
2. A feeling about the object: like or dislike component (Affective).
3. A tendency-towards-action: the action component (Behavioral).

It is important that we recognize that an attitude must be directed towards someone or something. We can hold a positive (or negative) attitude towards chemistry or the chemistry teacher, for example. The nature of attitudes can be expressed in a diagrammatic form (Figure 1.2).

One key feature of Figure 1.2 is the reminder that attitudes are hidden away in long-term memory and that they can only be measured indirectly. This is done by looking at behavior. However, this assumes that behavior is an accurate measure of the hidden attitude. In other words, it assumes that attitudes control behavior, and this is open to question. We shall return to this later.

Overall, it has been established that attitudes tend to be consistent and stable with time. Nonetheless, despite this stability, they are open to some change and development, although deeply held attitudes are highly inter-

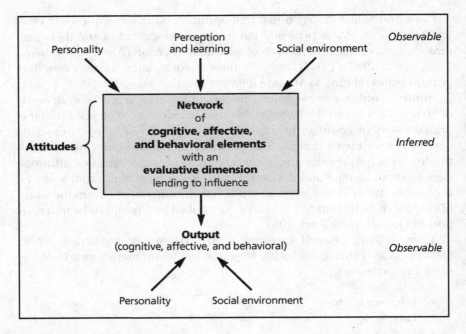

Figure 1.2 The nature of attitudes.

nalized and are resistant to modification (Johnstone & Reid, 1981). There is some evidence about the mechanisms by which attitudes do change and develop, relating the general stability of attitudes with the situations when change is possible (Johnstone & Reid, 1981; Reid, 1978, 1980). A textbook like *The Psychology of Attitudes* (Chaiken & Eagly, 1993) offers a thorough overview of the general findings from social psychology.

AN INTERIM SUMMARY

It is possible to summarize the key findings from social psychological research:

- An accepted description of an attitude is that it "is a psychological tendency that is expressed by evaluating a particular entity with some degree of favour or disfavour" (Eagly & Chaiken, 1993).
- Attitudes are stored in long-term memory and cannot be measured directly. They are described as latent constructs.
- Attitudes involve elements of knowledge, feeling, and behavior.
- Attitudes strongly influence behavior.

- Attitudes tend to show considerable consistency and are relatively stable.
- Attitudes are open to change and development, given the right circumstances.
- Attitudes enable humans to make sense of their world, themselves, and their relationships.
- There is considerable difficulty about their measurement.

ATTITUDES IN SCIENCE EDUCATION

In the field of science education, interest in attitudes was seen as important as the curriculum reform movements of the 1960s and 1970s were accompanied by falling enrolments in chemistry and physics in England and the U.S. Ormerod and Duckworth's (1975) excellent summary brought together much of this early work in a useful form, reflecting mainly the situation in England while Gardner (1975) also compiled a useful review.

My own country of Scotland did not experience this "swing from science" in the 1960s (Dainton, 1968). This may reflect the kinds of changes in the chemistry and physics curriculum that occurred then, these being very different from those in either England or the U.S. A much later study in Scotland looked at what attracted learners to physics, and this revealed the key factors very clearly (Reid & Skryabina, 2002a). The nature of the physics curriculum was found to be critical as well as the key importance of fully qualified, committed, and supportive teachers. Indeed, even today the three sciences and mathematics are the most popular elective subjects in Scotland at senior school level, judged by the uptakes, a position they have held for over 40 years (Scottish Qualification Authority, n.d.).

Despite the vast numbers of studies and the enormous amount of effort which was expended in trying to elucidate what was happening attitudinally as school pupils and university students elected *not* to study the physical sciences in many countries, clear insights did not emerge. Many have noted this with frustration. Some studies did offer limited new insights (e.g., Yilmaz, Boone, & Andersen, 2004), while others pinpointed interesting features that might be useful in curriculum planning (e.g., Myers, Boyes, & Stanisstreet, 2004). Some looked beyond school education (e.g., Rispoli & Hambler, 1999). However, a look at the detail of the Skryabina (2000) study shows that the methodology she adopted holds the key. The usual methods that science education researchers have taken from the world of social psychology are inappropriate and simply cannot offer the needed key insights.

A recent review Reid (2010) of the main findings related to attitudes towards the sciences summarized the key issues (Table 1.1).

TABLE 1.1   Key Research Findings Related to Attitudes

Research Finding	Implication
Interest develops early (by about age 14)	Expend our effort and energy with younger pupils.
Boys and girls are equally interested but areas of interest vary.	Ensure that themes and topics contain a balance of themes to interest both genders.
School teachers have a very critical role	Expend our effort and energy in supporting and resourcing teachers.
Things outside the school have almost no impact	Science Centres, one off events and network TV may be fun. They have very limited lasting impact.
There is a successful curriculum approach	This is based on the concept of applications-led curricula (Reid, 2000).
Integrated science courses are disasters	Chemistry needs to be taught by chemists, physics by physicists and biology by biologists at all secondary levels.
Career potential must be perceived	The considerable career openings for those with school qualifications in the sciences needs to be made explicit.

There are numerous specific studies, and some of these are summarized by Reid (2006, pp. 15–16). Looking at teaching and learning, Soyibo and Hudson (2000) report on attitudes in relation to computer-assisted instruction, and Thompson and Soyibo (2002) look at practical work. Sadly, both studies lose potentially interesting detail by the measurement techniques employed. Berg, Bergendahl, and Lundberg (2003) looked in detail at an expository approach compared to an open-ended approach in laboratory work. They avoided the methodological pitfalls of scaling and offered some very useful thoughts.

Brossand, Levenstein, and Bonney (2005) report an interesting study of attitudes in relation to a citizen science project. They found little attitude change but, almost certainly, the methodology used lost key detail. However, consistent with the findings of Skryabina (Reid & Skryabina, 2002a), they found that school pupils and university students claimed not to have been influenced towards the sciences by media events or by events like science festivals.

There are numerous studies looking at the interface between primary and secondary schools. Integrated science was starting to take hold as the norm in the early secondary years in the 1970s and, shortly after, primary science was seen as an important development. Although there are few studies, early work seems to show that integrated science does *not* improve attitudes relating to science (Brown & Davis, 1973; Brown, 1974), whereas considerable doubt was expressed in a later and somewhat broader review of the literature (Venville, Wallace, Rennie, & Mallone, 2002). Hadden and Johnstone (1982; 1983a,b) noted the very marked deterioration of attitudes in the early years of secondary, a deterioration that was far more marked in science in other areas of the curriculum. Similarly, primary teachers' confidence was found to be very low when asked to teach outside their area of competence (Harlen & Holroyd, 1997).

Other studies have looked at gender differences (e.g., Stark & Gray, 1999). Dawson (2000) looked at upper primary pupils attitudes while Reid and Skryabina (2002b) looked at gender across the age range from 10 to age 20 in relation to physics. In an interesting recent study, Spall, Barrett, Stanisstreet, Dickson, and Boyes (2003) focused on biology and physics undergraduates, although they did not emphasize the gender issues so much. Gauld and Hukins (1980) discuss many studies that tried to relate aspects of *scientific attitude* and various curriculum developments, mostly in the U.S.

ATTITUDE TARGETS

One of the important features of attitudes is the realization that there has to be an attitude "target." In science education, four areas of targets can be identified:

Attitudes towards:
  a. The science subject itself as a discipline
  b. The learning of the science subject (and perhaps learning more generally)
  c. Topics and themes covered in a particular course (e.g., themes of social awareness)
  d. The methods of science (the so-called scientific attitude)

By far the greatest area of interest has been that of attitudes towards chemistry and physics. Recent studies include Ramsden (1998); Osborne, Driver, and Simon (1998); Osborne, Jonathan, Simon, and Collins (2003); Bennett, Rollnick, Green, and White (2001); Pell and Jarvis (2001a,b, 2004); Krogh and Thomson (2005); Cleaves (2005). Some studies have looked at specific aspects. For example, Reid and Skryabina (2002a, 2002b) looked at physics while Shah (2004) focused specifically on laboratory work in chemistry.

There are a few studies which have looked at attitudes towards learning but most have focused on Higher Education. Selepeng (1999) looked at biology undergraduates; Mackenzie, Johnstone, and Brown (2003) looked at medical students, while Al-Shibli (2003) looked at how attitudes related to studies in the sciences developed over the final three years of school and the four years of a degree with Omani students. All of these studies followed the work of Perry (1999) and, while largely attitudinal in focus, there were wider dimensions.

Studies related to social awareness are very rare, although this area is assuming increasing importance in many curriculum specifications. The modern trend is to see this in relation to scientific literacy although there are few clear descriptions of what is meant by this term. Reid (1978, 1980) carried out some early work and showed how the curriculum could be presented to enable attitudes relating to the way chemistry was related to the world around could be developed fairly easily. This work was followed up by Al-Shibli (2003) who showed that the same approach worked in the development of attitudes towards learning with Omani students. A very recent study Chu (2008) used a similar approach in developing attitudes in relation to the genetics revolution. She worked with younger school students and found the same approach developed in the 1970s was also very effective in this area.

In looking at the fourth "target," some major problems are evident. There are major difficulties in describing what we mean by the "scientific attitude" in a way agreed by the majority although one attempt is described (Reid, 2006). Another problem lies in grasping whether this is really an attitude or a method of working (like a cognitive skill). One recent major study looked at scientific thinking (Reid & Serumola, 2006a,b), and this was followed up by another major study that took the ideas further and gained many useful insights (Al-Ahmadi, 2008). One paper has drawn in a small portion of the findings (Al-Ahmadi & Oraif, 2009).

Gauld and Hukins (1980) reviewed the area many years ago. They noted the lack of clarity and the lack of precision in the use of key words. Perhaps the problem really is that the "scientific attitude" is really a group of many attitudes. It also involves a high degree of skills character. The person has to be willing (based on attitude) but also has to be skilled (based on knowledge and experience).

ATTITUDES AND BEHAVIOR

Early work in social psychology reveals the considerable confusion over attitudes and behavior as well as attitudes and emotions. While attitudes can possess considerable emotional dimensions, attitudes are not emotions. While attitudes can be strongly influenced by past behavior and can lead to future behavior, attitudes are not the same as behavior. The confusions are compounded by the fact that the only way we can make any measurement of attitudes held by groups of people is to observe aspects of their behavior.

Because much of the work on attitudes has focused on attitudes *towards* the science (often physics or chemistry), the behavioral aspects have received much attention. The behavior here is reflected in whether the students want to continue with studies in chemistry or physics. Various models have tried, on the basis of empirical evidence, to show the extent to which attitudes influence subsequent behavior.

Ajzen and Fishbein (1980) developed a generally accepted model (theory of reasoned action). Their studies showed that behavior is rather well predicted from what they described as behavioral intentions (*are you intending to do... ?*). In turn, the behavioral intentions were predictable by attitudes towards the possible behavior and social norms (opinions of others). Later, the model was adapted and was called the *Theory of Planned Behavior* (Ajzen, 1985). It has to be stressed that Ajzen (1989) was only looking at planned behavior, behavior under the deliberate control of the person. The model developed by Ajzen can be illustrated in Figure 1.3. We can illustrate this specifically by looking at student deciding to study more physics (Figure 1.4).

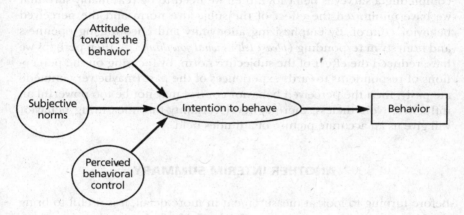

Figure 1.3   The Theory of Planned Behavior.

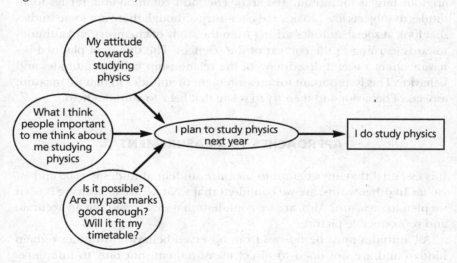

Figure 1.4   Illustrating the Theory of Planned Behavior.

Ajzen found that this model did account for much behavior and that the first factor (attitudes towards the behavior) was often the most important factor. The results from the work of Skryabina (2000) on attitudes to physics were found to fit this model.

The importance of the model in attitude measurements is considerable. We attempt to measure attitudes (hidden away in the long-term memory part of the brain) by measuring behavior. Behavior is *only* a good measure of attitudes if the effects of the subjective norm and the perceived behavioral control are minimized. This leads to our approaches to measurement. If we are going to attempt to measure attitudes by measuring behavior (and completing a survey is behavior), then we need to be reasonably sure that we have minimized the effect of the subjective norm and the perceived behavior control. By emphasizing anonymity and encouraging openness and honesty in responding (*please tell us what you think...*), then perhaps we have reduced the effect of the subjective norm. By focusing on the perceptions of respondents towards experiences of the past (maybe very immediate past), then the perceived behavior control may not be too powerful an influence. Nonetheless, there is never certainty that measuring behavior will give us an accurate picture of attitudes held.

ANOTHER INTERIM SUMMARY

Before turning to look at measurement in more detail, it is useful to bring the ideas discussed above together. In the field of science education, there are four targets for attitude research. The most common area relates to attitudes to subjects like physics and chemistry, although there are some studies that look at social attitudes arising from the study of chemistry and attitudes towards learning in the context of the sciences. The theory of planned behavior offers a useful description of the relationship between attitudes and behavior. This is important for measurement of attitudes in that we measure aspects of behavior and then try to relate that back to attitudes held.

APPROACHES TO MEASUREMENT

It is essential that any attempt to measure student attitudes is *valid* and *accurate*. In other words, are we confident that what we are measuring is what we plan to measure? Also, are we confident that we are gaining an accurate and reproducible picture?

All attitudes must be *inferred* from observed behavior. Attitudes remain hidden and are not open to direct measurement, but only to inference. This is not as serious as it sounds. We are happy to award grades and even degrees on the basis of marking samples of student work (behavior) and are able to make all kinds of statements with considerable confidence on learner understanding based on such evidence. How confident are we that our marks and grades actually reflect what is "hidden" in the brain of our students? We need to use caution in interpreting our examination results. Equally, we need to use caution when measuring the attitudinal.

In an important early discussion, Cook and Selltiz (1964) categorize the techniques of attitude measurement into five types. However, in schools and universities, paper-and-pencil techniques must dominate. Indeed, a little thought tells us that there are only three possible ways to approach attitude measurement in educational contexts:

- Set a written assessment (like a survey)
- Ask oral questions (like an interview)
- Observe some specific aspect of behavior (like behavior as they learn).

Let us look briefly at these three approaches.

Surveys

Surveys are socially acceptable and allow us to handle the large numbers in schools and universities. The response patterns obtained can give extremely accurate pictures of specific student attitudes. However, there is always an element of doubt whether responses reflect reality or aspiration (the reality–aspiration problem is identified by Danili, 2004).

Experience with self-report techniques (like the technique of Likert, 1932) shows that, under most conditions, respondents are remarkably honest and consistent in their responses (Clarkeburn et al., 2000; Reid, 2003, 2006). It is very clear that, given large samples and sensible test conditions, survey data can be extremely robust. The main problem rests with the way data from surveys are handled. Most studies have adopted approaches which are difficult to defend. This latter point we shall discuss in some detail later.

In looking at paper-and-pencil techniques, several types of question have been identified:

1. Those with a format similar to that developed by Osgood
2. Those with a format similar to that developed by Likert
3. Rating questions
4. Situational set questions

These approaches have been discussed briefly elsewhere, and the relative strengths and weaknesses have been summarized (Reid, 2003). In the semantic differential format developed by Osgood and colleagues (1957), the respondent is invited to place a check in one of a set of boxes placed between two adjectives, two adjectival phrases, or even two brief sentences. This defines both ends of the scale. In questions following the Likert format, respondents respond by showing a measure of agreement or disagreement. Thus, only one end of the scale is precisely specific, the other having to be deduced. Rating questions allow the respondents to select from a list, indicate preferences, or place items in a list in their own chosen order. Situational set questions are difficult to develop. In these, the respondent is placed in a simulated situation and reactions are invited. Examples are given in the Appendix.

Interviews

Interviews can often give fascinating insights and details not open to surveys. They can sometimes reveal surprising outcomes. However, they are very time-consuming and it is never easy to summarize a set of interviews down into a clear cut set of conclusions.

Interviews can be highly structured or totally open, but a very useful kind of interview can be described as semi-structured. Here the interviewer has a set of questions for discussion, but there is freedom to elaborate or move from the agenda as appropriate. With highly structured interview, data analysis can be more straightforward. For example, we can simply record the proportion of students who like laboratories, used a given textbook regularly, re-wrote their lecture notes after lectures, and so on. However, open interviews tend to be less structured and students can respond in widely different ways using widely different language. Indeed, in open interviews, the student may even determine the agenda. There is a considerable amount of research following this latter paradigm (e.g., Marton, 1986).

Behavior

Watching student behavior can also offer fascinating insights, as Johnstone and Wham found (1982) in an early study of student behavior in undergraduate laboratories, but it is also time-consuming. There is, inevitably, a measure of uncertainty that the observed behavior reflects attitudes in that it is not easy to control the subjective norms or perceived behavioral control.

VALIDITY AND RELIABILITY

Two of the problems in all educational measurement relate to validity and reliability. Validity is simply the extent to which the measurement actually measures what is intended. Thus, for example, if we wish to test student ability at, say, quadratic equations, then we ask the students to solve quadratic equations of varying complexity and see how successful they are. With attitudes, it is somewhat more complex. It is much more difficult to be certain that students are answering questions in a survey in such a way that their responses reflect underlying attitudes.

There are, however, several possible ways forward:

- Have all the questions to be asked checked by one or people who know the student population well
- Have all the questions to be asked checked by one or people who have some experience of the attitudes being explored
- Develop the questions using student language, perhaps following a discussion with another equivalent group of students in advance
- Interview a sample of the student group afterwards and explore why they responded in they way they did
- Interview a sample of the student group afterwards and to see whether their responses in the interviews are consistent with their responses in the survey

Reliability generates considerable controversy. Reliability explores the extent to which the responses are accurate. Suppose a student responded to a survey one day. Would the responses be similar the following day, or following week, for example? The problem is that there is enormous confusion over *internal consistency* and what we might call test-retest reliability. Test-retest reliability is what is important: we need to know if the measurement made one day will be very similar if made under similar conditions on another occasion. Gardner (1995) discusses this helpfully. Nonetheless, the literature is full of confusion in this area.

We need to stop and think for a moment. Suppose you wanted to measure my height. A measuring tape might be useful. It is clear you are trying to measure what you intend to measure. However, is the tape reliable? Will it stretch with repeated use? Is the scale marked on in the right units? There is no point in measuring my height with a tape that only has a scale marked to the nearest meter. Will my height be approximately the same on consecutive days? Will it be the same at different times of the day? There are many such simple questions, and we all know how to make a measurement of height that will reflect quite accurately what my height actually is. The best way may be to measure my height more than once.

Let me illustrate the utter confusion over internal consistency. For example, suppose a short survey asks ten questions about ten *different* aspects of learning school chemistry. Internal consistency is meaningless. The examiner does *not* set ten questions to test the same thing. The usual pattern is to set ten questions to test ten *different* topics, skills, or themes. The pupil performance in the questions may or may not correlate with each other. Internal consistency might offer evidence that, if a student knows one area, he might well perform well in another, but that does not say anything about the reliability of the test.

The literature is full of supposed attitude measures that quote statistics like split-half reliabilities and Cronbach's Alpha (Cronbach, 1951). These are measures of internal consistency. There is a complete lack of clear thought here. Suppose we have a survey with 20 questions. One common way is to correlate the responses of the odd numbered questions against the responses to the even numbered questions. First of all, correlation does NOT tell us that two measures are measuring the same thing. I could take 1000 men and measure their weight and height. The two measures will correlate highly but weight is NOT height! Gauld and Hukins (1980, p. 131) offer a useful comment when they say, "One can have conceptually distinct measures (like weight and height) which are empirically correlated and conceptually related measures which show almost no correlation." This is strong advocacy for caution.

Thinking again of attitude survey questions, taken to its logical absurdity, it is often argued that the higher the value of the correlation coefficient, the more reliable is the test. Imagine we obtain a coefficient of 1. Why ask twenty questions? Why not ask ten? Or even just one?

Cronbach's alpha is a measure that brings together all the inter-item correlations into one number. It is simply a measure of internal consistency. A simple analysis of the mathematics of this statistic shows how difficult it is to gain a low value! It says nothing about test-retest reliability. In education, Cronbach's alpha is not a very helpful statistic simply because so much is multi-dimensional.

Genuine reliability is really only assessed by using the questions on more than one occasion, but this is often not possible. However, where it has been done, all the evidence suggests that, with good samples, and the carefully controlled use of surveys, reliability is high. Thus totally undermines the oft-quoted comment that surveys tell us nothing. The inference from such a statement is that respondents tick boxes fairly randomly. Careful analyses have shown that this is usually not true at all (see Reid, 2006). However, it must be stressed that surveys tell us what respondents think. Surveys cannot measure things like learner characteristics with any degree of accuracy, as Hindal (2007) has demonstrated.

STATISTICAL ASPECTS

In many of the studies in attitude measurement, there is lack of understanding of the dangers of correlation. Indeed, in most cases, the wrong

TABLE 1.2   The Folly of Calculating Means

	Strongly Agree	Agree	Neutral	Disagree	Strongly Disagree	Mean Score
Group 1	13	25	44	57	19	2.7
Group 2	24	23	38	31	42	2.7

statistical method for correlation is used. Pearson correlation can *only* be used with integer data, approximately normally distributed. All survey data fail on the first point as the data are ordinal. Many survey questions fail on the latter, the data forming far from normal distributions. Kendall's Tau-b has to be used for correlation purposes, but you will rarely find this in most research studies.

Indeed, this is part of a wider problem with most survey data. It is very common to find researchers calculating means and standard deviations of responses to survey questions. It is *mathematical nonsense* to carry out such procedures with ordinal data. It is also very misleading. Here is a fictional example which illustrates this (Table 1.2).

In Table 1.2, two groups have completed the same question on a survey. The usual way is to allocate a value of 5 for "strongly agree," 4 for "agree," 3 for "neutral," 2 for "agree" and 1 for "strongly disagree." A moment's thought reveals the meaningless of this, but it is widely done. It is saying that a person who agrees with the statement has, in attitude terms, twice the value of a person who disagrees. This has no meaning at all! Indeed, the numbers 1 to 5 are ordinal. We could equally use A, B, C, D, E.

In the table, the mean score for the upper line is found by:

$$\frac{(13 \times 5) + (25 \times 4) + (44 \times 3) + (57 \times 2) + (19 \times 1)}{13 + 25 + 44 + 57 + 19} = \frac{430}{158} = 2.7$$

The lower line is calculated in the same way.

However, the two groups clearly are showing *very different* response patterns to this question but the totally invalid method used suggests that they are the same because they give the same mean.

Most studies make it even worse. They use the ordinal numbers of a set of questions and then add up the so-called *scores* for each person in all the questions to give a total score. This assumes that we can add ordinal numbers—a mathematical nonsense. It also assumes that the value of a number in one question is roughly the same as the value of the same number in another. This cannot be assumed, and there is *no* way of knowing.

Let us stop and think of a typical test in, say, chemistry. Let us imagine we have set a test with ten questions on ten different themes of the curriculum, each question worth ten marks. Imagine we develop a marking brief and we follow it as carefully as we can. This gives some indication of the accuracy of our marking, a part of test reliability. We then add up the marks each student has obtained in each question to give a total as a percentage. Is this a reasonable way forward?

Assuming we are reasonably experienced in test setting, then the ten questions may well be of similar difficulty. Thus, a mark in one question will be worth roughly the same as a mark in another. If the questions vary widely in difficulty, the problem is easily solved simply by standardizing the marks for each question before adding them up. This is a simple technique to ensure that the marks in all questions carry the same value. This can be done quickly using a spreadsheet. However, is it valid to *add* marks in one question that tests organic structures to the marks in another question testing, say, the reactions of chlorine? It will only be valid if we wish to gain a measure of the student's overall performance in the chemistry course we have taught and the skills associated with organic structures are of approximately equal value to the skills of the reactions of chlorine. This may well be true, but it depends on the curriculum we are teaching.

Now let us go back to an imaginary survey. Suppose we ask ten questions relating to student attitudes towards chemistry, and the students respond on a five-point scale, following the Likert format. First of all, there is no way of knowing that a "score" on one question is similar to a "score" on another. The numbers used are ordinal and *cannot* be studied in this way. Mathematically, ordinal numbers *cannot* be added or subtracted. Secondly, even if the first criterion were true, we can only add up the marks if the questions are asking things that, together, give us a meaningful measure (like the student's overall performance in the chemistry course we have taught, as above). The usual way this is checked by researchers is to use correlation. However, we have shown that correlation cannot be interpreted to say that two measures that correlate highly measure the same thing. Remember, height and weight correlate highly. They are obviously not the same.

However, there is a more powerful statistical technique that can be used to explore if the questions are measuring the same thing. This is often called Factor Analysis, and one method used widely today is known as Principal Components Analysis. Factor analysis looks to see if there are underlying factors that might rationalize the range of intercorrelations that are found when making several measurements.

To illustrate the technique, here is the analysis obtained in one study looking at something completely different. In a major study with a sample of 809 senior school students in the United Arab Emirates (Al-Ahmadi, 2008, page 145), the marks for the national examinations in biology, chemistry, physics, and mathematics were obtained. The students also completed a test of physics understanding and a test that aimed to measure scientific thinking. A principal components analysis, using varimax rotation, was carried out on the data. The Scree plot indicated that there were three components, and this accounted for over 90% of the variance. Possible names for the three components were suggested. Table 1.3 shows the loadings obtained: Loadings can be thought of as the correlations between the actual measurements and the underlying factors. High loadings are emboldened for clarity (Table 1.3).

This table needs a little interpretation. The numbers known as loadings can be seen as the correlations between each measurement and the underlying components. What it tells us is that the four examinations were testing the same skill and, after looking at the examination papers, Al-Ahmadi identifies this as recall/recognition. The test of physics understanding was not testing this, and she deduced that it might well actually be testing physics understanding. The test that aimed to measure scientific thinking was not testing recall/recognition, nor was it testing the same as the physics understanding test.

The aim in this analysis was to explore the test that had been designed to measure scientific thinking and which was based on physics content. It was possible that this test might simply be testing physics knowledge or physics understanding. The factor analysis eliminated this possibility in that the loadings of the test of scientific thinking are extremely low on these components.

In another study that involved 40 questions looking at a range of attitudes, a principal components analysis, using varimax rotation, was carried out on the data (Oraif, 2007). The Scree plot suggested that there was no simple factor structure underlying the pattern of inter-question correlations (which had been done correctly using Kendall's Tau-b). Indeed, it took 12 components to account for only 55% of the variance and this % of the variance is too low to be acceptable.

I have analyzed the data from large numbers of surveys from all over the world. I have never found one where the principal components analysis sup-

TABLE 1.3 Factor Loadings (N = 809)

	Component 1	Component 2	Component 3
Possible names >>>>>>	Recall and recognition	Scientific Thinking Skills	Physics Understanding
Test of Scientific Thinking	0.15	**0.98**	0.12
Test of Physics Understanding	0.13	0.12	**0.99**
Physics Examination	**0.91**	0.16	0.05
Chemistry Examination	**0.93**	0.06	0.09
Biology Examination	**0.93**	0.13	0.14
Mathematics Examination	**0.88**	0.08	0.10

Source: After Al-Ahmadi, 2008, p. 145.

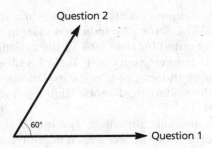

Figure 1.5 The meaning of correlation.

ported a factor structure. However, most attitude studies do not use any form of factor analysis. Even when they do, the loadings, which are far too small, are often accepted. You will recall that loadings are simply the correlations of the various questions with the underlying factors. They must be *very high*. To see what is acceptable, we need to think further about correlations.

Suppose we have two questions and the responses of our students give a correlation of, say, 0.5 between these two questions. We can represent one question as measuring along a line. The other question then can be represented by a line at an angle of 60°. The correlation value, mathematically, is the *cosine of the angle* between the two lines and the cosine of 60° is 0.5 (Figure 1.5).

In much work using survey questions, obtaining a correlation between two questions is often used to suggest that the two questions are measuring the same thing and therefore outcomes can be added. First of all, correlation does not, of itself, guarantee that two measurements are measuring the same thing. Secondly, the two lines in Figure 1.5 are in very different directions and that suggests that they are *not* measuring the same thing. Try to measure my height by holding the measuring tape at 60° to the vertical. You would end up making me a giant—over 3.5 meters tall.

It takes a correlation of over 0.7 to reduce the angle to 45° and even a correlation of 0.9 still gives an angle of approximately 25°. Yet, in many attitude studies, loadings of 0.4 are taken as acceptable. This reveals a major weakness in many studies.

EXPERIMENTAL ERROR

Another problem rests with the nature of the measuring instrument. If a student checks a box one to the left or one to the right of the correct estimation of his view, there is an error of ±20% on a five-point scale. We say that data from attitude surveys is "soft." This is in contrast to the types of data we obtain from reading a voltmeter or a burette. Here, the measuring errors are very small. The data are said to be "hard." It does not matter how much clever statistics we apply to soft data, the data are still soft. The use of statistics cannot alter that.

This leads to an important principle.

> With the present state of knowledge, it is not appropriate to use surveys to measure attitudes for individual students. The potential errors in responses are relatively high, and the typical item scale only has five or six points on it. It is rather like measuring a student's performance in a physics examination using a scale where students either get 10%, 30%, 50%, 70%, or 90%.

There is another important principle:

> In the present state of knowledge, attitudes *cannot be measured in any absolute sense*, with any degree of certainty.

This leads to the key way attitude surveys can be used:

> Responses to attitude measures can be *compared*: before and after some experience, between two different groups such as male and female. However, there must be large numbers in all samples.

ATTITUDE SCALING TECHNIQUES

This is the most common approach used today. It was developed by taking an approach used in psychology and applying it in education contexts. The usual way is to use the Likert style of question but it applies equally to the semantic differential questions developed by Osgood and colleagues (2003). Here is an outline of what is often done.

Suppose we want to assess our students' attitudes about a laboratory course they have just undertaken. We develop a set of statements relating to this course. Here is a possible set which was used by a researcher when looking at student reactions to chemistry laboratory experiences. Originally, they were part of a longer survey and were used with first year Scottish undergraduate students (Shah, 2004). Later, they were used with various postgraduate groups in Pakistan (Shah, Riffat, & Reid, 2007).

Think about your experiences in laboratory work you have experienced in chemistry.
Tick the box which best reflects your opinion.

	strongly agree	agree	neutral	disagree	strongly disagree
(a) I believe that the laboratory is a vital part in learning chemistry	☐	☐	☐	☐	☐
(b) I prefer to have written instructions for experiments.	☐	☐	☐	☐	☐
(c) All the chemicals and equipment that I needed were easily located	☐	☐	☐	☐	☐
(d) I was unsure about what was expected of me in writing up my experiment	☐	☐	☐	☐	☐
(e) Laboratory work helps my understanding of chemistry topics	☐	☐	☐	☐	☐
(f) Discussions in the laboratory enhance my understanding	☐	☐	☐	☐	☐
(g) I only understood the experiment when I started to write about it afterwards	☐	☐	☐	☐	☐
(h) I had few opportunities to plan my experiments	☐	☐	☐	☐	☐
(i) I felt confident in carrying out the experiments in chemistry	☐	☐	☐	☐	☐
(j) I found writing up about experiments pointless	☐	☐	☐	☐	☐
(k) The experimental procedure was clearly explained in the instructions given	☐	☐	☐	☐	☐
(l) I was so confused in the laboratory that I ended up following the instructions without understanding what I was doing	☐	☐	☐	☐	☐
(m) I feel that school examinations should take account of laboratory experiments I have completed	☐	☐	☐	☐	☐

Scaling might have scored each item as follows: "strongly agree" = 5, "agree" = 4, "neutral" = 3, "disagree" = 2 and "strongly disagree" = 1. The scores for all the items are then added up (reversing as necessary) to give a to-tal score. If a researcher was being a bit more careful, he would have run a factor analysis on the responses to check if there was some common underlying construct that rationalized the pattern of responses. In other words, is there evidence that there is some common idea (for example, attitude to chemistry laboratories) that could be used to interpret the responses? Often, questions that loaded onto this construct at 0.6 (say) or better would be selected and the other questions rejected. Happily, Shah had the wisdom not to do any of this.

The whole methodology of scaling is completely flawed:

1. We have no way of ensuring that steps on the scale in any question are equally spaced. It is impossible to measure the spacing.

2. Values on one question may not be comparable to those on another. Almost certainly, they are not and we have no way of knowing the relative values.
3. Correlations often assume normality, which is frequently absent. Indeed, large skewing may be "desirable" and polarization may often be observed. Factor analysis relies on correlation.
4. Correlations do not necessarily imply direct relationship. Correlations certainly cannot be used, on their own, to suggest that two questions are measuring the same thing. Loadings are simply correlations with the supposed underlying construct.
5. Similar total scores may be obtained for very different patterns of attitudes.
6. There are problems associated with errors and "softness" of this kind of ordinal data. For an individual, this may be as large as ±20% in any question
7. Combining scores hides the rich detail arising from each question. The distribution in EACH separate question is what matters.

Let us look at some of this a little further. Imagine two students who respond to the set of questions above. Here are the outcomes (Table 1.4).

TABLE 1.4 Response Patterns

	Student 1						Student 2					
	strongly agree	agree	neutral	disagree	strongly disagree	'Score'	strongly agree	agree	neutral	disagree	strongly disagree	'Score'
	5	4	3	2	1		5	4	3	2	1	
(a)		X				4		X				4
(b)	X					5			X			3
(c)				X		2		X				4
(d)		X				4				X		2
(e)					X	1		X				4
(f)				X		2	X					5
(g)	X					5				X		2
(h)		X				4			X			3
(i)				X		2	X					5
(j)	X					5					X	1
(k)		X				4					X	1
(l)			X			3		X				4
(m)				X		2	X					5
					Total	**43**					**Total**	**43**

The response patterns in Table 4.1 are fictional. However, the two students achieve the same '*score*' in the set of questions but it is very clear that they do *not* hold the same views relating to laboratories. In the original research, Shah wisely looked at each question in turn and did not use scaling The pattern of responses from several hundred students offered very useful information and pinpointed *precisely* the features of the student laboratory experience which were perceived as helpful and which needed to be re-thought. This is useful information to direct future planning. A '*score*' is meaningless. Indeed, a '*score*' is incapable of revealing those aspects of the student laboratory experience which are perceived as positive or otherwise.

Look at four of the questions again:

a. I believe that the laboratory is a vital part in learning chemistry
e. Laboratory work helps my understanding of chemistry topics
f. Discussions in the laboratory enhance my understanding
i. I felt confident in carrying out the experiments in chemistry.

The responses to these questions correlate with each other. Correlation merely shows that those who were most positive on one question tend to be most positive on another. Adding up scores produces a fairly meaningless number simply because the questions are asking *different* things.

Let us go back to the earlier work of psychologists. They were tending to look at broad constructs, like attitudes about race or political attitudes. The work of psychologists often depended heavily on pre-testing and carrying out correlations between the responses on the questions. If correlations were reasonably high, it was assumed that the questions were measuring similar things. That is quite an assumption, and Gardner (1995, 1996) has usefully discussed some of the issues involved here.

In the earlier work of psychologists, they tended to look at large groups and see if they differed in the mean scores obtained. Because of the nature of what they were exploring, the method, although fundamentally flawed, still revealed when groups were *very* different. In educational research, the nature of the attitudes being explored is almost always highly multi-faceted or multi-dimensional. It is not possible or helpful to reduce this to a single "score," simply because we lose all the important detail that can guide us in attempting to make the learning experience more fruitful.

VALIDITY AND RELIABILITY

The ideas of validity and reliability have already been discussed briefly. There is also the question about how *accurate* a set of questions is. If responses are on a five- or six-point scale (which is quite typical), potential error limits are very large. It is amazing that means and standard deviations are often quoted to two or more decimal places! It is not right to take a mean of ordinal data and, even if it was, the result must be quoted to the level of accuracy of the input data. The data simply are soft and must be treated with due caution.

There have been all kinds of ingenious attempts to use different statistical techniques on data gathered in attitude surveys. They all have to face the simple problem that we are looking at soft data. No matter how sophisticated are the techniques we use, statistics cannot remove this feature. Nor can it remove the fact that the data from surveys are ordinal in nature. These data must be handled with statistics appropriate to such data. Indeed, given the nature of the data and their softness, the fewer statistics used the better. We are in danger in deluding ourselves that sophisticated techniques can give us numbers with spurious accuracy.

An example illustrates the problem. Suppose we wish to gain some insight into our health and fitness. We book into a clinic where all kinds of measurements are made: blood pressure, temperature, heart rate, heart rate when exercising, height, weight, cholesterol level, subcutaneous fat levels, and so on. We come away with a printout of all these measurements. It would be folly to add up the data on our print-out! We cannot add weight to blood pressure to heart rate and so on. The "sum" would be utterly meaningless (Reid, 2003). Much of this was discussed long ago by Johnstone (1982) but the publication was not circulated widely. Sadly, his advice has not been followed by the majority.

SUMMARY

- Most attitudes related to science education are *multi-faceted* and any attempt to reduce measurement to a final score for each individual will tend to give a meaningless number.
- The data obtained from questionnaire questions are ordinal in nature and attention must be paid to the nature of these data in determining what statistical techniques should be used. Ordinal numbers cannot be added, subtracted, multiplied or divided.
- The data obtained from questionnaire questions are *soft* in the sense that, for an individual, error limits are high. Any "score" obtained for an individual will, therefore, be open to considerable error.
- In general, parametric statistics should only be used with caution in any attitude measurements in that the conditions for the use of such techniques are often not satisfied.

- Reliability cannot be measured by traditional methods like "split half" and Cronbach's Alpha: These measure consistency, which may not even be desirable.
- Principal Components Analysis must be used with caution. It may rely on assumptions of approximate normality. While it is unlikely to lead to wrong conclusions, it may end up reducing the data in such a way that rich detail is lost. Nonetheless, it can give useful insights.
- Attitude scaling should *never* be used for logical and statistical reasons. It must be recognized that such an approach will only show gross trends simply because the rich detail is lost in the adding process, and precision is lost because the method relies on the application of scale methodologies to ordinal data and the use of inappropriate statistics.

EXAMPLES

The sad thing is that the literature is replete with examples of the kind of careless analysis being discussed here. It is only possible to add marks from separate questions if there is some evidence that the marks are on a *comparable scale*, measure *something meaningful* and give a total that carries *some clear meaning*. In attitude work, this is rarely possible. Paper after paper has developed some "instrument" or "tool." Often, factor analysis data are presented. Sometimes, the set of questions is applied to two discrete groups and great confidence is expressed when they differ in their mean scores. Thus, those who are physics majors are found to give a higher mean score than those who are having to take physics as an outside subject. This is utterly pointless for two main reasons. Firstly, any teacher could tell us this outcome—it is completely to be expected. Secondly, the research study fails to offer any insights about *why*. What is it about physics that makes it attractive to learners? This is where the work of Skyrabina (2000) stands out. At the end of her study, she could pinpoint precisely what features of study in physics (at all levels) were critical in making learners enthusiastic about physics. This is vital information, for it can inform curriculum planners.

Osborne and colleagues (2003) make some interesting observations when looking at falling numbers choosing the physical sciences in England and Wales. They stressed the sheer number of sub-constructs underlying attitudes to a science and they are aware of the relative nature of some attitude measurement approaches. Of great interest is their statement that "Attitude scales...while they are useful in identifying the problem, they have been of little help in understanding it" (Osborne et al., 2003, p. 1059). This is, again, almost certainly because the use of such scaling obscures and loses rich detail which could offer such help.

Krogh and Thomson (2005) also note that there are many attitude studies but few conclusions. This is consistent with Schebeci (1984), who stated, "It is disappointing that the set of conclusions which can be drawn from such a large body of literature is so limited" (p. 46). Gardner (1995) hints at part of the answer—methodology. Sadly, Krogh and Thomson (2005) still use flawed methodology.

Jarvis and Pell (2004) use a scaling approach and, amazingly, quote means and standard deviations of outcomes from a Likert scale to two decimal places, with a relatively small sample. In this report, they find very little change—almost certainly another example of the adding up process losing interesting detail. However, all kinds of interesting and useful observations are made, presumably based on specific questions but lost in the data in table 14 (pp. 1804–1805). Similar weaknesses are to be found in Tuan, Chin, and Shieh (2005) while many of these weaknesses are happily avoided in Jenkins and Nelson (2005).

While Gogolin and Swartz (1992) identify many of the problems, they still proceed to develop a scale for looking at attitudes to biology. Their study is a nice example of high intercorrelations between different items which are, nonetheless, asking very different questions: an example of the weakness of depending on correlation. They followed the usual procedure of scoring, and gaining total scores but observed how small were the changes in the means of the total scores. This is almost certainly an example of numbers cancelling each other out, with concomitant loss of important detail. As a result, their conclusion tells us nothing new: "The results of this study . . . suggest that attitudes toward science change with exposure to science but that the direction of change may be related to the quality of that exposure" (Gogolin & Swartz, 1992, p. 500).

Fraser and Wilkinson (1993) looked at Science Laboratory Classroom Climate and present a massive piece of work. With huge samples involving six countries, they developed a scale with several subscales and subjected the data to rigorous analyses. Despite all this rigor, the problems still remain. Responses are scored on a five-point scale. Correlation was used as an indirect tool, and loadings from the factor analysis were accepted at levels as low as 0.3 to 0.4. The problem is that most of the interesting detail (which can be used in practical terms to inform decision taking) was lost in inappropriate statistical manipulations.

Let us look at three of their questions, purporting to measure the same variable:

Students in this laboratory class get along well as a group.
Students have little chance to get to know each other in this laboratory class.
Members of this laboratory class help one another.

Loadings on what they call a "social cohesiveness" dimension are quoted as 0.70, 0.49, and 0.67 respectively. However, it is obvious that these three questions are measuring very different things. Indeed, a correlation of 0.7 geometrically corresponds to an angle of just over 60°. This hardly suggests measuring in the same direction.

It is the responses to the questions illustrated above, each taken on their own, that are really interesting. The first question is indeed getting at social cohesiveness of the class, while question 2 really reflects on organization within and beyond the class. Question 3 is quite different again. It might throw light on the moral atmosphere in this particular population, or it might offer insights into the way the class was run: collaboratively or competitively. No doubt other interpretations could be offered for all three. However, it is these insights *from individual questions* that are really valuable and will lead to better laboratory experiences, and this is the laudable aim implicit in this paper. Large amounts of data are being handled in a way that has lost the important insights that can inform curriculum planners and course designers.

Some 25 years ago, Schebeci (1984) reviewed much work available at that time. He noted the many surveys that have shown that most studies are psychometrically unsound. He was aware of the criticisms of attitude scaling but did not pursue these. He is also critical of much work, and he noted, "Given established methods, why do researchers continue to report studies in which attitude instruments are used which are either clearly invalid or which few data on reliability or validity are reported?" (p. 43). He expressed surprise that, despite the large number of reported studies, there was great difficulty in establishing what were the key variables affecting attitudes. It is a pity that he did not suggest that one of the reasons might be that the scaling methodologies were *incapable* of producing evidence about the key variables since so much rich detail was lost in the desire to produce a number.

WAYS FORWARD

In the discussion so far, I have emphasized much that is wrong in the ways attitude measurement has been approached in education and, specifically, science education. All this seem very negative. However, there are many excellent studies in the literature, studies that have not used inappropriate methodologies.

The first things is to ask *why* we want to measure attitudes in the context of science education. Despite papers that sometimes suggest otherwise, there is no evidence that poor attitudes towards the sciences are *causing* students to turn away from studying the sciences. It seems much more likely that poor attitudes towards the sciences are caused by the way the sciences are often presented at school stages, the work of Skryabina (2000) showing this perhaps most clearly. This is not usually the fault of teachers but arises from bad curriculum design, overloaded curricula, and inappropriate assessment.

Therefore, the idea of measuring attitudes towards the sciences (or a specific science) to find those whose attitudes are poor with some intention of directing help at them is doomed to failure from the outset. The answer is already known. If we change the curriculum in line with the evidence from research, then attitudes may well improve. Two monographs exist that have summarized the research evidence relating to curriculum construction (Mbjiorigu & Reid, 2006a,b). In simple terms, negative attitudes towards the studying of the sciences (usually chemistry and physics) are caused by inappropriate curricula. If we change the curricula, then it is likely that attitudes will be more positive. A summary of the findings of the monograph on chemistry is shown in the appendix.

Perhaps there are three broad research areas where further work is important. These apply to both school and university education. Again, it is apparent that details are needed and the approach of attitude scaling simply will not give us the information we need.

- It will be important to measure the attitudes of students towards a science being studied so that we can see what *specific* aspects of their learning experiences are perceived in a positive light and what are causing problems.
- It will be important to measure the attitudes of students towards learning in the sciences to identify those *specific* features of the learning processes which are viewed negatively. This may well follow the analyses of Perry (1999) although his work can be seen in a much wider learning context.
- It will be important to measure the attitudes of students towards the very large range of social issues that relate to developments in the sciences. This will identify whether our courses are equipping learners to be able to discuss some of the key issues of modern life, using arguments that are informed and balanced: themes like the genetics revolution and its impact on medicine and life decisions, nuclear energy, resource depletion, pollution issues, global warming, and so on.

While many studies continue to use highly flawed approaches, there are numerous excellent examples of better ways forward in the literature. For example, a very insightful set of papers was written by Hadden and Johnstone (1982; 1983a,b). These looked at attitudes towards sciences as they formed at primary school stages, the rapid deterioration of attitudes in the early years of secondary school (now known to be almost certainly caused by integrating science), and how attitudes were determinants of decisions to pursue science-based courses or not.

A similar methodology was employed by Reid (1980) and Reid and Skrya-bina (2002a, 2002b). The first study looked at social attitudes related to chemistry and pinpointed specific areas where attitude development was possible through a curriculum intervention. The latter study looks at attitudes towards physics among students aged 10 to age 20. Interestingly, this study revealed considerable similarities with the much earlier Hadden and Johnstone studies, along with some key differences. It was possible to relate these features *precisely* to the kind of learning the learners were experiencing.

Encouragingly, the literature has some examples of other ways forward. Perhaps, the answer lies in reversing the typical procedure. Instead of *adding* the responses in questions to give a score that is then analyzed, the answer lies in analyzing each question and then "qualitatively adding" the outcomes obtained. This approach has been used in the study by Al-Shibli (2003). He used an 18-item questionnaire, with seven populations of different ages, and his study considered attitudes toward *learning* in the context of science subjects. His samples were enormous (>2000). The study was interested in the way attitudes to learning changed over the seven-year educational experience, and the questions were derived from previous work and extensive pre-testing. Four areas of attitudes were explored: the role of the learner, the role of the teacher, the nature of knowledge, and the role of assessment (following ideas from Perry, 1999). The distribution of responses in each of the eighteen questions for each of the seven age groups was summarized and comparison made between successive year groups using the chi-square statistic.

He then looked at each question in turn and he found that there were patterns of statistical differences and patterns of actual responses across questions in each area which offered a rich insight into what was happening in this particular educational journey. The precision of this approach enabled him to pinpoint the need to develop new curriculum experiences to enrich *specific* experiences of the students. Attitude scaling would have lost all this detail and would have made the pinpointing of curriculum 'gaps' more or less impossible.

There are numerous studies which have looked at laboratory work. For example, Johnstone, Watt, and Zaman (1998) looked at attitudes related to undergraduate physics laboratories. Interestingly, this study showed quite remarkable attitude movements. Each question was analyzed in turn and offered very valuable and clear cut insights.

In an important contribution, Bennett et al. (2001, p. 834) noted seven areas of contention in attitude measurement:

1. Lack of precision over key definitions of terms
2. Poor design of instruments and individual response items within instruments
3. Failure to address matters of reliability and validity appropriately

4. Inappropriate analysis and interpretation of data
5. Lack of standardization of instruments
6. Failure to draw on ideas from psychological theory
7. Failure to formulate the research with reference to the theory of data collection tools.

Gardner (1995) had identified many of these previously. However, Bennett and colleagues (2001) describe how they attempted to minimize the problems. They came up with an ingenious approach. Using a small sample to generate the reasons, they developed items that looked like:

I like it when the lecturer gives us small tasks to do in lectures.
   A   I AGREE with this statement because it improves my understanding.
   B   I AGREE with this statement because it improves my concentration.
   C   I AGREE with this statement because I learn better in a group.
   D   I DISAGREE with this statement because discussions are for tutorials.
   E   I DISAGREE with this statement because it increases the noise and wastes time.
   F   I DISAGREE with this statement because in a big class some students do not participate.
   G   None of these statements fits my view, which is . . .

Their approach is a good example in that the validity is high simply because the reasons offered were derived from students themselves. They suggest that the data can be presented simply as sets of frequencies. However, they tried to get a score for each student using a method derived from that used by Thurstone in 1929. Although this approach reduces some of the problems in handling the data, it still runs the risk of losing rich and insightful detail. They almost seem to admit this when they note the narrow range of total scores obtained. Nonetheless, the paper describes ways forward that have potential.

CONCLUSIONS

The following is an attempt to bring the evidence together to offer a constructive way forward that may assist future work. There are important general principles.

Attitude development is very important in that it will influence future behavior, and such behavior may have very significant consequences for the individual and for society. We might all agree that, as teachers, we are aiming to generate a future population who are equipped in one or more of the sciences at some level and, more importantly, can see the nature and place of the sciences as an integral element in our modern development as a society.

The needs of research in science education require approaches that can offer accurate, rich, and detailed insights into how attitudes develop with time and relate to the learning experiences in our schools and university classes. In all of this, we need to recognize that, with current knowledge, *absolute* measures of attitudes are impossible. Only comparisons can be made. We also need to recognize that any attempt to measure attitudes for individual learners is also impossible and, probably, unethical. We can measure patterns of attitudes across populations of learners and relate specific aspects to specific features of the learning processes.

Most attitudes related to science education can be described as highly *multi-faceted* or *multi-dimensional*. Any attempt to reduce measurement to a final score for each individual will tend to give a meaningless number and lose important detail.

There are numerous paper-and-pencil approaches based on the structures developed by Likert (1932), Osgood and colleagues (1957) as well as rating questions and situational set questions (Reid, 2003). As all these types of questions have different strengths and weaknesses, it is recommended that surveys incorporate more than one format. Of course, interviews can offer useful insights, and it is suggested that these can be used alongside surveys. Sometimes, a number of interviews can define the agenda for a major survey. On other occasions, interviews can explore issues raised by a previous survey.

In considering surveys, I am often asked about sample size. There are no neat answers to this question, but here is a simple rule of thumb. In surveys, it is best to compare groups using a statistic like chi-square. Most statistical approaches are highly sensitive when the size of each sample group is about 200. Thus, if we wish to look at the attitudes of boys and girls to a new course in physics that we have introduced, then surveying 200 boys and 200 girls is ideal. However, in one school, this is probably impossible. Either we need to involve several schools (and this makes any findings much more generalizable), or we might be able to get round this by carrying out the survey with two successive year groups in one school. The question is how small the sample can be to obtain useful insights. With samples below 100, chi-square becomes difficult to apply in that there are category limits for each category of responses. When samples are nearer 50, the sensitivity of most statistical tests falls away and drawing conclusions may well be difficult. Overall, aim for 200 for each sample group, but anything between 100 and 200 may well give most valuable insights.

RECOMMENDATIONS

This chapter has aimed to set the scene in relation to attitude measurement in science education. The nature, purpose, and importance of attitudes have been outlined. The main purpose is to look critically at the methods for attitude measurement that have been adopted by many in the field of science education. These have too often been taken uncritically from approaches in psychology where the agenda is very different.

It has been noted that the huge number of studies already reported have not really given the science education community a clear set of guidelines. This is simply because the methodologies used have been inappropriate. Indeed, they are not only inappropriate and incapable of giving the answers we need. They are simply wrong. The methods break so many fundamental rules of mathematics and statistics that it is frightening. The key recommendations in using surveys are offered in Table 1.5.

TABLE 1.5   Some Recommendations

The data obtained from survey questions are ordinal in nature and attention must be paid to the nature of these data in determining what statistical techniques should be used.

The data obtained from survey questions often do not follow any specific pattern of distribution. Statistics which assume an approximation to a normal distribution must be used with great care.

The data obtained from survey questions are 'soft' in the sense that, for an individual, errors limits are high. Any method of analysis which relies on individual responses to individual question is, therefore, highly suspect

Principal Components Analysis must be interpreted with caution. While it is unlikely to lead to wrong conclusions, it may end up reducing the data in such a way that rich detail is lost. Nonetheless, it can give most valuable insights.

Reliability cannot be measured by traditional methods like 'split half' and Cronbach's Alpha: these measure consistency which may not even be desirable. Reliability means test-retest reliability and sensible use of surveys will ensure reasonable reliability

Reliability can be ensured by doing things like using large samples, careful pre-testing, checking that test conditions are socially acceptable, using enough questions with cross checks (e.g., repeated questions, similar questions).

Attitude scaling should never be used. It must be recognised that such an approach will only show gross trends simply because the rich detail is lost in the adding process and precision is lost because the method relies on the application of scale methodologies to ordinal data and the use of inappropriate statistics.

The pattern of responses in each question should be analysed separately.

The pattern of responses to compare year groups, genders, or 'before and after' measurements can be analysed by the appropriate use of chi-square ('goodness of fit' or 'contingency' test, depending on the research question asked).

If inter-item correlation is used, then Kendal's Tau-b correlation must be employed. It is unlikely that Pearson correlation will give totally wrong results but significance levels may be misleading.

Validity of surveys can be checked by numerous approaches, including seeking opinions of a group of those who know the population, using questions derived from the population, comparing conclusions with other evidence. Validity is critical and an invalid test will always be unreliable.

On this basis, attitude scaling is to be rejected. New approaches (some well documented in the literature for several years) are recommended. The aim in all studies related to attitudes is to be able to give clear guidance to curriculum planners and to teachers at all levels so that positive attitudes towards the sciences and their study can be enhanced on the basis of sound empirical evidence. In addition, there is a major need for looking at attitudes related to the themes and methods of science, two areas of vital importance in generating an educated society.

The aim in all attitude research is to give answers. Answers are needed by policy makers, curriculum planners, and teacher trainers as well as by teachers themselves. Sadly, many studies have used flawed methodologies and have not focused on the key issues that are of importance in science education today. Look at many papers: Do they tell us anything useful that can bring insights of benefit to learners?

The aim of all science education is to educate in and through the science disciplines. For success, we need committed and enthusiastic learners who hold positive attitudes towards the science studied as well as informed attitudes in relation to the themes covered. Attitude research needs to keep these aims in sharp focus and then, using correct methodologies that accept the multi-dimensional nature of such attitudes, much can be offered to enhance the learning of future generations

REFERENCES

Ajzen, I. (1985). From intentions to actions: a theory of planned behavior. In J Kuhl & J Beckman (Eds.), *Action control: From cognition to behavior.* New York: Springer.

Ajzen, I. (1989). *Attitude personality and behaviour.* Chicago: Dorsey.

Ajzen, I. & Fishbein, M. (1980). *Understanding attitudes and predicting social behavior.* Englewood Cliffs, NJ: Prentice Hall.

Al-Ahmadi, F.M. (2008). *The development of scientific thinking with senior school physics students.* PhD Thesis, University of Glasgow, Glasgow. Available online at: http://eleanor.lib.gla.ac.uk/search/a

Al-Ahmadi, F.M. & Oraif, F. (2009). Working memory capacity, confidence and scientific thinking, *Research in Science and Technological Education, 27*(2), 225–243.

Al-Shibli, A.A.S. (2003). *A Study of Science Teachers Perceptions of Learning in the Education Colleges in the Sultanate of Oman,* PhD Thesis, University Glasgow, Glasgow.

Allport, G.W. (1935). Attitudes. In C.M. Murchison (Ed.), *Handbook of Social Psychology* (pp. 798–844). London: OUP.

Bagozzi, R. P. & Burnkrant, R. E. (1979). Attitude organisation and the attitude-behaviour relationship, *Journal of Personality and Social Psychology, 37,* 913–929.

Bennett, J., Rollnick, M, Green, G., & White, M. (2001). The development and use of an instrument to assess students' attitude to the study of chemistry. *International Journal of Science Education, 23*(8), 833–845.

Berg, C.A.R., Bergendahl, V.C.B., & Lundberg, B.K.S. (2003). Benefitting from an open-ended experiment? A comparison of attitudes to, and outcomes of, an expository versus open-ended inquiry version of the same experiment. *International Journal of Science Education*, 25(3), 351–372.

Brossand, D., Levenstein, B., & Bonney, R. (2005). Scientific knowledge and attitude change: the impact of a citizen science project, *International Journal of Science Education*, 27(9), 1099–1121.

Brown, S.A. (1974). *Integrated Science–A Useful Innovation?*, Conference Proceediungs, ASE Regional Meeting, Aberdeen, 10th April, 1974.

Brown, S.A. & Davis, T.N. (1973). Attitude goals in secondary school science. *Scottish Educational Studies*, 5, 85–94.

Chu, Y-C. (2008). *Learning difficulties in genetics and the development of related attitudes in Taiwanese junior high schools*. PhD Thesis, University of Glasgow, Glasgow.

Clarkeburn, H., Beaumont, E., Downie, R., & Reid, N. (2000). Teaching biology students transferable skills. *Journal of Biological Education*, 34(3), 133–137.

Cleaves, A. (2005). The formation of science choices in secondary school. *International Journal of Science Education*, 27(4), 471–486.

Cronbach, L.J. (1951). Coefficient Alpha and the internal structure of tests, *Psychometrika*, 16, 297–334.

Cook, S.W. & Selltiz, C. (1964). A multiple indicator approach to attitude measurement. *Psychological Bulletin*, 62, 36–55.

Dainton, F.S. (1968). *The Dainton Report: An inquiry into the flow of candidates into science and technology*. London: HMSO.

Danili, E. (2004). *A study of assessment formats and cognitive styles related to school chemistry*. PhD Thesis, University of Glasgow, Glasgow.

Dawson, C., (2000). Upper primary boys' and girls' interests in science: have they changed since 1980? *International Journal of Science Education*, 22(6), 557–570.

Doob, L.W. (1947). The behavior of attitudes. *Psychological Review*, 54,135–156.

Eagly, A. H. & Chaiken, S. (1993). *The psychology of attitudes*. London: Harcourt Brace Jovanovich.

Festinger, L.A. (1957). *A theory of cognitive dissonance*. Stanford, CA: Stanford University Press.

Fraser, B. J. & Wilkinson, D. (1993). Science laboratory climate in British schools and universities. *Research in Science and Technological Education*, 11(1), 49–70.

Gardner, P.L. (1975). Attitudes to science: A review. *Studies in Science Education*, 2, 1–41.

Gardner, P.L. (1995). Measuring attitudes to science: Unidimensionality and internal consistency revisited. *Research in Science Education*, 25(3), 283–289.

Gardner, P.L. (1996). The dimensionality of attitude scales. *International Journal of Science Educatiion*, 18(8), 913–919.

Gauld, C.F. & Hukins, A.A. (1980). Scientific attitudes: A review. *Studies in Science Education*, 7, 129–161.

Gogolin, L. & Swartz, F. (1992). A quantitative and qualitative inquiry into attitudes towards science of non-science college students. *Journal of Research in Science Teaching*, 29(5), 487–504.

Hadden, R. A. & Johnstone, A.H. (1982). Primary school pupils' attitude to science: The years of formation. *European Journal of Science Education*, 4(4), 397–407.

Hadden, R. A. & Johnstone, A.H. (1983a). Secondary school pupils' attitude to science: The year of erosion. *European Journal of Science Education, 5*(3), 309–318.

Hadden, R.'A. & Johnstone, A.H. (1983b). Secondary school pupils' attitude to science: The year of decision. *European Journal of Science Education, 5*(4), 429–438.

Harlen, W. & Holroyd, C. (1997). Primary teachers' understanding of concepts of science: Impact on confidence and teaching. *International Journal of Science Education, 19*(1), 93–105.

Heider, F. (1944). Social perception and phenomenal causality, *Psychology Review, 51,* 358–374.

Heise, D.R. (1970). The semantic differential and attitude research. In G. F. Summers (Ed.), *Attitude measurement* (pp. 235–253). Chicago: Rand McNally.

Hindal, H.S. (2007). *Cognitive characteristics of students in middle schools in State of Kuwait, with emphasis on high achievement.* PhD Thesis, University Glasgow, Glasgow.

Hovland, C.I., et al. (1953). *Communication and persuasion.* New Haven, CT: Yale University Press.

Jarvis, T. & Pell A. (2004). Primary teachers' changing attitudes and cognition duirng a two-year science inservice programme and their effect on pupils. *International Journal of Science Education, 26*(14), 1787–1811.

Jenkins, E.W. & Nelson, N.W. (2005). Important but not for me: Students attitudes towards secondary school science in England. *Research in Science and Technological Education, 23,* 41–57.

Johnstone, A.H. (1982). Attitude measurements in chemistry: Pitfalls and pointers. In *Chemical education research—Implications for teaching* (pp. 90–103). London: Royal Society of Chemistry.

Johnstone, A.H. & Reid, N. (1981). Towards a model for attitude change. *European Journal of Science Education, 3*(2), 205–212.

Johnstone, A.H. & Wham, A.J.B. (1982). Demands of practical work. *Education in Chemistry, 19*(3), 71–73.

Johnstone, A.H., Watt, A. & Zaman, T.U. (1998). The students' attitude and cognition change to a physics laboratory. *Physics Education, 33*(1), 22–29.

Katz, D. & Sarnoff, I. (1954). Attitude change and structure. *Journal of Abnormal and Social Psychology, 49,* 115–154.

Krech, D. & Crutchfield, R.S. (1948). *Theory and problems of social psychology.* New York: McGraw-Hill.

Krogh, L. B. & Thomson, P. V. (2005). Studying students' attitudes towards science from a cultural perspective but with quantitative methodology: Border crossing into the physics classroom. *International Journal of Science Education, 27*(3) 281–302.

Likert, R. (1932). A technique for the measurement of attitudes. *Archives of Psychology, 140,* 5–55.

Mackenzie, A.M., Johnstone, A.H. & Brown, R.I.F. (2003). Learning from problem based learning. *University Chemistry Education, 7,* 13–26.

McGuire, W. J. (1985). Attitudes and attitude change. *Advances in Experimental Social Psychology, 16,* 1–47.

Marton, F. (1986). Phenomenography: A research approach to investigating different understandings of reality, *Journal of Thought, 21,* 28–49.

Mbajiorgu, N. & Reid, N. (2006a). *Factors influencing curriculum development in chemistry.* A Physical Sciences Practice Guide, The Higher Education Academy, Hull.

Mbajiorgu, N. & Reid, N. (2006b). *Factors influencing curriculum development in physics.* A Physical Sciences Practice Guide, The Higher Education Academy, Hull.

Myers, G., Boyes, E., & Stanisstreet, M. (2004), School students' ideas about air pollution: Knowledge and attitudes. *International Journal of Science Education,* 22(2), 133–152.

Oraif, F. A. (2007). *An exploration of confidence related to formal learning in Saudi Arabia.* PhD Thesis, University of Glasgow, Glasgow.

Ormerod, M. B. & Duckworth, D. (1975). *Pupils' attitudes to science.* Windsor, UK: NFER.

Osborne, J., Simon, S., & Collins, S., (2003). Attitudes towards science: a review of the literature and its implications, *International Journal of Science Education,* 25(9), 1049–1079.

Osborne, J., Driver, R., & Simon, S. (1998). Attitudes to science: Issues and concerns. *School Science Review, 79*(288), 27–33.

Osgood, C.E., Suci, C.J., & Tannenbaum, P.H. (1957). *The measurement of meaning.* Urbana, IL: University of Illinois Press.

Pell, T. & Jarvis, T. (2001a). Developing attitude to science scales for use with children of ages from five to eleven years. *International Journal of Science Education,* 23(8), 847–862.

Pell, T. & Jarvis, T. (2001b). Developing attitude to science scales for use with primary teachers. *International Journal of Science Education,* 25(10), 1273–1295.

Pell, T. & Jarvis, T. (2004). Primary teachers changing attitudes and cognition during a two-year science in-service programme and their effect on pupils. *International Journal of Science Education, 26*(14), 1787–1811.

Perry, W.G. (1999). *Forms of ethical and intellectual development in the college years: A scheme.* San Francisco: Jossey-Bass.

Ramsden, J. (1998). Mission impossible?: Can anything be done about attitudes to science? *International Journal of Science Education, 20*(2), 125–137.

Reid, N. (1978). Simulations and games in the teaching of chemistry. *Perspectives in Academic Gaming and Simulations,* Vol 1 and 2, 92–97.

Reid, N. (1980). Simulation techniques in secondary education: Affective outcomes. *Simulation & Gaming, 11*(1), 107–120.

Reid, N. (2000). The presentation of chemistry: Logically driven or applications led? *Chemistry Education: Research and Practice, 1*(3), 381–392.

Reid, N. (2003). *Getting started in pedagogical research.* Higher Education Physical Sciences Practice Guide, Hull, Higher Education Academy.

Reid, N. (2006). Thoughts on attitude measurement. *Research in Science and Technological Education, 24*(1), 3–27.

Reid, N. (2010). A scientific approach to the teaching of chemistry. *Chemistry in Action, 90.* University of Limerick, Ireland.

Reid, N. & Serumola, L. (2006a). Scientific enquiry: The nature and place of experimentation: A review. *Journal of Science Education, 7*(1), 1–15.

Reid, N. & Serumola, L. (2006b). Scientific enquiry: The nature and place of experimentation: Some recent evidence. *Journal of Science Education, 7*(2), 88–94.

Reid, N. & Skryabina, E. (2002a). Attitudes towards physics. *Research in Science and Technological Education, 20*(1), 67–81.

Reid, N. & Skryabina, E. (2002b). Gender and physics, 2002. *International Journal of Science Education, 25*(4), 509–536.

Rhine, R.J. (1958). A concept formation approach to attitude acquisition. *Psychological Review, 65*, 362–370.

Rispoli, D. & Hambler, C. (1999). Attitudes to wetland restoration in Oxfordshire and Cambridgeshire. *International Journal of Science Education, 21*(5), 467–484.

Roediger, H. L., Ruston, J. P., Capaldi, E. D., & Paris, S. G. (1984). *Introduction to psychology*. Boston: Little, Brown.

Scottish Qualifications Authority. (n.d.). *Annual Reports, 1962–2009* . Dalkeith, Scotland: Author. Recent reports online at: http://www.sqa.org.uk/sqa/3773.html

Schebeci, R.A. (1984). Attitudes to science: An update. *Studies in Science Education, 11*, 26–59.

Shah, I. (2004). *Making university laboratory work in chemistry more effective*. PhD Thesis, University of Glasgow

Shah, I., Riffat, O. & Reid, N. (2007). Students' perceptions of laboratory work in chemistry at school and university in Pakistan. *Journal of Science Education, 8*(2), 75–78.

Selepeng, D. (1999). *An investigation of the intellectual growth in undergraduate biology students using the Perry scheme*. PhD Thesis, University of Glasgow.

Soyibo, K. & Hudson, A. (2000). Effects of computer-assisted instruction (CAI) on 11th graders' attitudes to Biology and CAI and understanding of reproduction in plants and animals. *International Journal of Science Education, 18*(2), 191–200.

Spall, K., Barrett, S., Stanisstreet, M, Dickson, D., & Boyes, E. (2003). Undergraduates views about biology and physics. *International Journal of Science Education, 21*(2), 193–208.

Skyrabina, E. (2000). *Students' attitudes to learning physics at school and university levels in Scotland*. PhD Thesis., University of Glasgow, Glasgow.

Stark, R. & Gray, D. (1999). Gender preferences in learning science. *International Journal of Science Education, 21*(6), 633–643.

Thompson, J., & Soyibo, K. (2002). Effects of lecture, teacher demonstrations, discussion and practical work on 10th graders' attitudes to chemistry and understanding of electrolysis. *Research in Science and Technological Education, 20*(1), 25–37.

Thurstone, L.L. (1929). Attitudes can be measured. *Psychological Review, 36*, 222–241.

Thurstone, L.L. & Chave, E.J. (1929). *The measurement of attitudes*. Chicago: University of Chicago Press.

Tuan, H-L., Chin, C-C, & Shieh, S-H. (2005). The development of a questionnaire to measure student motivation towards science learning, *International Journal of Science Education, 27*(6), 639–654.

Venville, G., Wallace, J, Rennie, L., & Malone, J. (2002). Curriculum integration. *Studies in Science Education, 37*, 43–83.

Yilmaz, O., Boone, W.J., & Andersen, H.O. (2004). Views of elementary and middle school Turkish students towards environmental issues. *International Journal of Science Education, 26*(12), 1527–1546.

APPENDIX: EXEMPLAR SURVEY QUESTIONS

(1) What are your opinions about University Physics?
Place a tick in one box between each phrase to show your opinions.

I feel I am coping well	☐ ☐ ☐ ☐ ☐ ☐	I feel I am not coping well
I am not enjoying the subject	☐ ☐ ☐ ☐ ☐ ☐	I am enjoying the subject
I have found the subject easy	☐ ☐ ☐ ☐ ☐ ☐	I found the subject hard
I am growing intellectually	☐ ☐ ☐ ☐ ☐ ☐	I am not growing intellectually
I am not obtaining new skills	☐ ☐ ☐ ☐ ☐ ☐	I am obtaining new skills
I am enjoying practical work	☐ ☐ ☐ ☐ ☐ ☐	I am not enjoying practical work
I am getting worse at the subject	☐ ☐ ☐ ☐ ☐ ☐	I am getting better at the subject
It is definitely 'my' subject	☐ ☐ ☐ ☐ ☐ ☐	I am wasting my time in this subject

This set of eight measurements looks at different aspects of university physics and the measurements analysed separately. The style follows the model developed by Osgood et al. (1957).

(2) How do you see biology?
Please show your opinion by ticking one box on each line

	Strongly Agree	Agree	Neutral	Disagree	Strongly Disagree
(a) It is essential that every pupil learns some biology.	☐	☐	☐	☐	☐
(b) I should like to study more science (biology) in high school or university.	☐	☐	☐	☐	☐
(c) Using a textbook is more useful when you study than using your notebook.	☐	☐	☐	☐	☐
(d) Biology is a subject to be memorised.	☐	☐	☐	☐	☐
(e) Knowledge of biology is useful making world decisions.	☐	☐	☐	☐	☐
(f) Doing an examination in biology is stressful.	☐	☐	☐	☐	☐
(g) Biology is related my life.	☐	☐	☐	☐	☐
(h) Biology is useful for my career.	☐	☐	☐	☐	☐

This question follows the style developed by Likert (1932). It is unusual in that there are few studies relating to biology. Each item was analysed separately.

(3) Here are two other questions to illustrate other approaches which have been used successfully.

Here are several reasons why laboratory work is part of most chemistry course
Place a tick against the THREE reasons which YOU think are the most important.

Chemistry is a practical subject ☐
Experiments illustrate theory for me ☐
New discoveries are made by means of experiments ☐
Experiments allow me to find out about how materials behave ☐
Experiments teach me chemistry ☐
Experimental skills can be gained in the laboratory ☐
Experiments assist me to planning and organise ☐
Experimental work allows me to think about chemistry ☐
Experimental work makes chemistry more enjoyable for me ☐
Laboratory work allows me to test out ideas ☐

This question was looking to see how students saw the purposes of laboratory work. The frequencies of ticks in the boxes gave a pattern which allowed the researcher to determine the order of importance of the various reasons for laboratory work in chemistry. This, in turn, offered insights into why they held certain attitudes in relation to laboratory work

(4) Which factor(s) influenced your choice of planned honours subject(s)?
Tick as many as you wish.

☐ Enjoyment of subject   ☐ Friends
☐ Good grades at school in subject   ☐ Likely career opportunities
☐ Your teacher at school   ☐ Demonstrations, exhibitions, festivals
☐ Your parents   ☐ Any other factors (please list below)
☐ Information from mass media   _____

This question was seeking to explore which factors had influenced students towards physics and related subjects and gave a clear picture of three factors which dominated.

APPENDIX B: CURRICULUM DESIGN

The chemistry curriculum at school level should:

1. Be designed to meet the needs of the majority of pupils who will never become chemists (or even scientists), seeking to educate *through* chemistry as well as *in* chemistry.
2. Be strongly "applications-led" in its construction, the applications being related to the lifestyle of the pupils and being used to define the curriculum: fundamentally, the content is determined *not* by the logic of chemistry but by the needs of pupils.
3. Reflect attempts to answer questions like: What are the questions that chemistry asks? How does chemistry obtain its answers? How does this chemistry relate to life?
4. Not be too "content-laden," so that there is adequate time to pursue misconceptions, to aim at deep understanding of ideas rather than content coverage, and to develop the appreciation of chemistry as a major influence on lifestyle and social progress; avoid using analogies or models (or multiple models) in a way that causes information overload.
5. Not introduce sub-micro and symbolic ideas too soon or too rapidly; avoid developing topics with high information demand before the underpinning ideas are adequately established to overload and confusion.
6. Be set in language that is accessible to the pupils, avoiding the use of unnecessary jargon and offering careful clarification of words where the normal contextual meaning can cause confusion.
7. Be couched in terms of aims that seek to develop conceptual understanding rather than recall of information, being aware of likely alternative conceptions and misconceptions.
8. Offer experiences of graded problem-solving situations starting from the more algorithmic and moving on to the more open-ended.
9. Involve laboratory work with very clear aims: These should emphasize the role of lab work in making chemistry real as well as developing (or challenging) ideas rather than a focus on practical hands-on skills; lab work should offer opportunities for genuine problem solving.
10. Require assessment that is integrated into the curriculum and reflects curriculum purpose, is formative as well as summative, and aims to give credit for understanding rather than recall, for thinking rather than memorization.

As an extra, the curriculum should be taught by teachers who are qualified as chemists and are committed to the place of the discipline in its social context. This is an important factor in the developing of a soundly taught pupil population with positive attitudes towards chemistry. This has manpower and resource implications as well as implications for pre-service and in-service support (Mbajiorgu & Reid, 2006a).

CHAPTER 2

NEW APPROACHES TO THE STUDY OF STUDENTS' RESPONSE TO SCIENCE

Lars Brian Krogh
Center for Science Education, Aarhus University

At the very least, further research into attitudes has something to offer
by way of possible explanations for the persisting problem of the apparent
alienation of young people from science
—Ramsden, 1998, p. 134

INTRODUCTION

Attitudes towards science have been investigated for decades, and considering the "widely felt concern that there is a problem with students' interest in studying science, technology, engineering, and mathematics" (Osborne, Simon, & Tytler, 2009, p. 2), the conditions for a progressing research program seem to be present. The nature of attitudinal research, and its accumulated body of insights are evidenced in a series of reviews of the field (Gardner, 1975; Osborne, Simon, & Collins, 2003; Osborne et al., 2009; Schibeci, 1984). The discipline has evolved since Schibeci (1984) found reason to summarize, "It is disappointing that the set of conclusions which

45

can be drawn from such a large body of literature is so limited" (p. 46), but still more recent reviewers have identified various methodological aspects of attitudinal research that tend to undermine a more cumulative effort in the field. Consistently, reviewers bemoan the imprecision of the key-term *attitudes towards science* (e.g., Ramsden, 1998; Osborne et al., 2003). Osborne and colleagues state this explicitly: "30 years of research into this topic has been bedevilled by a lack of clarity about the concept of investigation" (Osborne et al., 2003, p. 1053) and go on to list a number of sub-constructs used more or less synonymously for attitudes towards science (p. 1054). Gardner (1995) has identified methodological flaws, in particular a frequent confusion between reliability and unidimensionality of scales invented to measure attitude. Several authors (Gardner, 1975; Ramsden, 1998; Schibeci, 1984) call for more "theoretical underpinning" and use of "theoretical frameworks" to inform attitudinal research. Finally, all reviews of attitudinal studies witness that research within this area has been and remains dominated by more or less post-positivistic perspectives and quantitative methods (e.g., Krogh, 2006; Osborne et al., 2003). Some limitations of this methodological preference have been indicated (Krogh & Thomsen, 2005; Osborne et al., 2003). However, in order to establish a rationale for the present work, I will take the discussion of inherent limitations of conventional approaches to attitudinal research a little further:

- *The lack of dialogue and synergy between conventional approaches* and *interpretive/naturalistic approaches to students' engagement with science.*

Husén describes the generally acknowledged

> conflict between two main paradigms employed in researching educational problems. The one is modeled on the natural sciences with an emphasis on empirical quantifiable observations which lend themselves to analyses by means of mathematical tools. The task of research is to establish causal relationships, to explain (Erklären). The other paradigm is derived from the humanities with an emphasis on holistic and qualitative and interpretive approaches (Verstehen). (Husén, 1997, p. 17)

Within the study of students' engagement with science this divide is present as a lack of shared discourses, limited exchange of results, and no deliberate division of labor between researchers from the conventional paradigm and researchers from the interpretive paradigm, predominantly researchers drawing on cultural perspectives. Manifestations of this unfruitful situation could be: the term "attitude" is absent from the SAGE Handbook of Qualitative Research (Denzin & Lincoln, 2005), while a rich body of interpretive studies of students' responses to science is absent from recent reviews of attitudes towards science (Osborne et al., 2003; Osborne et al., 2009). Osborne and colleagues implicitly address the need to add Verstehen/interpretive approaches to the repertoire of attitudinal research when writing: "Conventional quantitative instruments are useful in identifying the nature of the problem, [but] they have been of little help in understanding it" (Osborne et al., 2003, p. 1059). The strength of conventional approaches is to establish patterns of relationships between attitudes and a range of explanatory variables (Erklären)—their weakness is that they leave the *meanings* and *motives* (Verstehen) of such relations to speculation. Statistical measures provide little understanding of the unique motive structure of individual students, which actually guides their intentions, choices, and behavior related to school science and science careers. In other words, much could be achieved if Erklären and Verstehen-positions were brought into dialogue, despite traditional claims to incompatibility of underlying research paradigms (e.g., Walker & Evers, 1997). Placing attitudinal research within a contemporary pragmatic paradigm (Johnson & Onwuegbuzie, 2004) might be one way of overcoming these reservations (while keeping away from poststructuralism; see Guba & Lincoln, 2005). Mixed methods approaches where larger quantitative studies ("postpositivist methodology") are complemented with rich qualitative small sample studies (interpretive paradigm) would be the natural and useful consequence. If only some sharing of research objectives, discourse and theoretical perspectives could be achieved.

- *The need for entering cultural perspectives into "attitudinal" research to accommodate "new datasets" from international, comparative studies* (Osborne et al., 2009). Findings from the ROSE-study (Schreiner, 2006) and other international studies indicate that students' attitudes towards science are shaped as much by students' participation in late modern culture as by school science itself. Osborne and colleagues speculate that if "any explanatory hypothesis lies in the values of contemporary culture in which western youth is situated where future employment (at least until now) has not been a major threat and where some of the issues of identity discussed previously are more to the fore" (2009, pp. 11–12). Theorists of contemporary youth mentalities (Giddens, 1991; Ziehe, 2004) would suggest that the age of late modernity induces rapid changes in youth social characters, as well as deteriorate traditional authorities like schools, teachers, and science. From this perspective negative attitudes towards science may be ascribed to school science's inability to adapt quickly enough to the changing mentalities of adolescents. Furthermore, several of the explanatory hypotheses offered for girls'/women's lack of engagement with science (Blickenstaff, 2005) emphasize cultural gender biases that originate outside the science classroom. Such cultural interpretation suggests that students bring their life-worlds into school science, and it considers students' capacity to negotiate cultural differences as crucial for their attitudes towards science and science related career aspirations. Attitudinal theory needs a theoretical framework eliciting such cultural perspectives, as well as it needs to extend the present focus from school science classrooms to the interplay of classroom factors with both structural and personal variables.

- *Conventional studies tend to miss the developmental as well as the situational.* It has been emphasized that studies of students' development of attitudes towards science tend to be cross-sectional, and that real longitudinal studies are scarce (e.g., Ramsden, 1998). As longitudinal studies are time-consuming, this may appear as a practical delimitation of insights into how individual attitudes in the ST field develop. However, more principled limitations are inherent in conventional attitudinal research, as may be argued with the widely accepted conceptual framework for attitudinal research, the Theory of Planned Behavior of Ajzen (1985). Here, it is emphasized that attitude is a relation between person and some external object or specific behavior. Since school science is clearly another object than science as an enterprise at the societal level, it is no surprise that attitudes towards school science can be negative, while attitudes towards science as an enterprise for most students tend to be positive. Similarly, it is no surprise that attitudes towards school science may be related to but distinct from attitudes towards pursuing a ST career. The problem is that to describe the unitary developmental process where a student receives information about science and school science, responds, learns to value these, and ends up pursuing a ST career him/herself, one has to apply at least three different attitudes (towards science, school science, and science careers)—with no rules of how to correspond those. So, to understand the most vital process of internalization, there is definitely a need for a unifying framework. Some have suggested an identity framework to obtain an integrated description (e.g., Brickhouse & Potter, 2001; Carlone, 2005), but in this contribution I will outline how the cultural perspectives introduced by Phelan (Phelan, Davidson, & Cao, 1991), Costa (1995), and Aikenhead (1996) can be used for the same purpose. Another methodological issue with conventional attitudinal approaches is that it is very difficult to distinguish students' development processes from considerations of construct stability of attitudes and measurement reliability. Finally, conventional instruments for attitudinal measurement are build around items with a considerable degree of de-contextualization, written as they are far from the specific classroom setting and situated learning activities. Students' responses have to be constructed across a variety of situations (e.g., "We cover interesting topics in science class" from Simpson–Troost Attitude Questionnaire). In contrast to the conceptions of interest afforded by Germans Hidi, Krapp & Renninger (1992), conventional attitudinal research is short of a notion of situated attitudes. This may be due to the "erroneous perception that attitudes are stable and unrelated to cognitive states" (Ramsden, 1998, p. 131).

Summing it up, there is a need to enter interpretive and cultural perspectives to the field of attitudinal research, and to elaborate a theoretical framework that enables studies of developmental processes as well as more situated aspects. The aim of the present work is to provide first steps to this effort, describing two studies that address these issues differently and with different relative weights.

Entering Cultural Perspectives

In a series of interpretive studies, a particular Cultural Border Crossing perspective on students' response to schools and school science has been elaborated (Aikenhead, 1996; Aikenhead, 2001; Costa, 1995; Phelan et al., 1991). Based on a conception of culture by anthropologist C. Geertz (1973), Phelan and colleagues explicated culture as "norms, values, beliefs, expectations, and conventional actions of a group" (1991, p. 228), a definition that has been adopted by subsequent studies. Further, these studies share a multiple world view of students—that is, as participants in the worlds of families, world of peers, as well as the subcultures of schools and school science. The first study of Phelan and colleagues empirically investigated "how meanings and understandings derived from these worlds combined to affect students' engagement with schools and learning" (Phelan et al., 1991, p. 224). The pivotal finding was that perceived cultural borders between students' life-worlds and their local school culture influenced students' engagement with schools. The concept of cultural borders/boundaries was here defined as "real or perceived lines or barriers between worlds" (Phelan et al., 1991 p. 225). From thick descriptions of 54 high school students' multiple worlds and school trajectories, a typology was established relating the smoothness of students' transitions into school culture to the degree of congruence between this world and their other worlds: (1) congruent worlds/smooth transitions, (2) different worlds/boundary crossings managed, (3) different worlds/boundary crossings hazardous, and (4) borders impenetrable/boundary crossings insurmountable. The two categories of "different worlds" somehow reflected student agency and diversity: students would differ in their ability to negotiate given boundaries and the results accordingly differ. Costa (1995) extended this work to the field of school *science*, finding that students' response to school science could be understood in terms of a *two*-dimensional set of congruencies between students' life-worlds, and the worlds of schools and school science, respectively. Costa, however, introduced five student types to represent typical students' responses. This in effect turned the typology of cultural border crossings into a typology of students (see Table 2.1, which adds information on prevalent cultural processes from Jegede & Aikenhead, 1999).

Patterns of student responses are expressed through characteristic cultural processes and student type descriptors. Jegede and Aikenhead (1999) were first to relate different cultural processes to each student type: for example, congruent worlds make transition to school science easy and open for *enculturation*. Enculturation evokes the S&T careers aspirations typical for Costa's *Potential Scientists*. Conversely, when the students' worlds of family, peers and other schooling are at odds with the world of school science they are more likely forced into *assimilation* of the school science subculture, with instruction marginalizing their other sub-cultural belongings. As a result, students may be alienated from science, they may develop passive resistance mechanisms or they may invent non-meaningful coping strategies; all responses that allow them to get by without really entering the subculture of school science. The other student types represent different ways and degrees of engagement with science and can be used to anchor

TABLE 2.1 Student Types, Cultural Borders, and Cultural Processes

Life–world congruence with school	Life–world congruence with school science	Ease of transition	Student type	Cultural process
congruent	congruent	smooth	potential scientist	enculturation
congruent	inconsistent	managed	other smart kids	'anthropological'
inconsistent	inconsistent	hazardous	I don't know students	autonomous acculturation/ assimilation
discordant	discordant	impossible	outsiders	autonomous acculturation/ assimilation
irreconcilable	potentially compatible	frustratingly difficult (Aikenhead, 2001)	inside outsiders	autonomous acculturation/ assimiliation

speculations about who would be the realistic 'second tier' for recruitment to science. Most empirical cultural border crossing studies have focused on groups of students who are lacking in the science trail, and who should be expected to perceive hazardous cultural border crossings in relation to the white, western, male-flavored school science. However, the previous focus on ethnic/indigenous/non-western students and/or girls (Brown, 2004; Carlone, 2004; Chinn, 2002; Gilbert & Yerrick, 2001; Waldrip & Taylor, 1999) can be extended, for example, to accommodate the "new datasets" emphasizing culturally shaped youth mentalities (Osborne et al., 2009).

The cultural border crossing framework balances structure and agency in the description of students' response to science. Semantically, and throughout the literature, cultural border crossing is used to denote either a perceived "insurmountable" *structure* to climb (e.g., Phelan et al.'s definition above) or the very *process* of climbing/passing a border (e.g., Aikenhead, 2001). Structural aspects concern research questions like *What cultural borders are there?* (Erklären/Verstehen); *What meanings are ascribed to them?* (Verstehen); and *Which cultural borders are typically significant and have the greatest effect?* (Erklären). Some of the most important process aspects relate to questions like *How are the cultural borders actualized within science classroom?* (Verstehen) and *How do individual students negotiate?* (Verstehen). We see how the double nature of the concept encompasses both *Erklären* and *Verstehen* components. Consequently, the framework has previously unrealized potential to combine such perspectives and it invites to mixed methods research and the merging of knowledge forms.

First Approach: Exploring the Relation Between Attitudes Towards Science and Various Types of Cultural Border Crossings

In a previous study (Krogh & Thomsen, 2005), we established a framework for the use of CBCs in conventional attitudinal research. The overall aim of the study was to bridge traditional paradigms by exploring the following questions:

Q1: How can the concept of CBCs be used to add a cultural perspective to quantitative studies of attitudes towards science and choice of A-level subjects?

Q2: What types of CBCs turn out to be important, and how important are they compared with personality and classroom variables?

For the purpose of this chapter I will be focusing on elements of methodology.

The Relation Between Attitudes and CBCs

The studies of Phelan and Costa had related students' response to science to the amount and character of CBCs ("taxonomy"). Taking "attitude towards scienc" to be a special case of "response to science" we *hypothesized* that CBCs would influence attitudes, for example, that more hazardous CBCs tend to produce less favorable attitudes. The framework allows a closer empirical inspection of this relation.

A Typology of CBCs

Here, we were inspired by the founding father of social science M. Weber, (Weber, 1947) who analyzed human action in terms of four categories. According to Weber, purposeful *goal-rational action* is associated with both rational goals *and* means, while *value-oriented action* is striving for a substantive goal with rational means. *Affective action* is anchored in the emotional state of the actor rather than in the rational weighting of means and ends, and the final category, *tradition-bound action*, captures the great bulk of everyday action, to which people have become habitually accustomed. CBCs as barriers or processes mediating human action can be analyzed in closely related categories:

- Goal-rational border crossings: cognitive and instrumentalist by nature
- Value-oriented border crossings: cognitive, but non-instrumentalist by nature
- Affective border crossings: non-cognitive, emotional by nature
- Tradition-bound border crossings: non-cognitive, non-emotional, embodied and habitual by nature.

Specific CBCs of Different Types—Examples from an Exploratory Study

For explorative studies like the present one, the idea is to use the whole range of research studies, conventional as well as cultural, to track down and formulate CBCs of potential significance. From conventional studies in the Danish context we drew a few pre-validated scales (Humanist/Science Perception of Knowledge, Purpose of Knowledge Acquisition—see Zeuner & Linde, 1997) to represent CBCs, while cultural studies inspired us to the rest. Our concern was to have important CBCs of each type, and to allow variation in the formulation and operationalization of CBCs.

To fit in with a simple quantitative methodology all CBCs were embedded in single items with discrete response formats (Likert scales or similar, see Table 2.2).[1] Each useful item had the quality that certain students' responses could be interpreted as expressing a structural CBC. Sample items of each type, and some indication of the diversity of operationalizations are given in Table 2.2.

TABLE 2.2   Sample of Cultural Border Crossing Items of Various Types

CBC-type	Sample-items	Interpretation/Comment
Goal-rational CBC	*"What area would you choose for further education, if you were to choose right now?"* (list of options, including *science* or *technology*—each exemplified with professions)	If neither *science* nor *technology* is chosen by students, this is considered an indication of a Goal-rational/instrumentalist barrier to students' engagement with school science.
Value-oriented CBC	*"One must try to make knowledge as independent from human values as possible. Otherwise it cannot be trusted."* (Sample item from Scientific Perception of Knowledge scale: 3-item-scale, each item answered on 4-point scales from *completely agree* to *completely disagree*)	Low score on the SPoK scale is taken as an indicator of incongruence with scientific perceptions of knowledge. Such incongruence with the epistemological nature of school science constitutes a cultural border for students to manage.
Affective CBC	*"I consider it a problem that there is so little room for emotions in physics"* (4-point Likert scale with responses from *completely agree* to *completely disagree*)	High scores on this (cognitive formulation of an) affective item indicate an affective CBC; students with such a perception will have to struggle to find a place for their own feelings in the science classroom.
Tradition-bound CBC	*"Does your father have an education or occupation related to science or technology? (e.g., engineer, laboratory technician, computer scientist, scientist, veterinarian)."* (similar question for mother's occupation. Yes/No responses only)	If students answer no to both of these questions, science and technology are unlikely to be significant parts of their family background, and their personal habitus. Again this would constitute a cultural barrier to their enculturation to science.

Analytical Procedures

Exploring Where the Most Frequent CBCs Are

Once the correspondences between students' responses and specific CBCs had been established, simple descriptive statistics provided a useful first picture of the potential barriers. Figure 2.1 shows the coarse dichotomized picture of students' responses[2] to a pool of single-item potential CBCs: Are they perceiving this CBC or not? More sophisticated measures might be used, but the benefits are not obvious here.

A number of features are immediately noticed: students' other worlds (world of family, friends etc.) explicitly entered through highly frequent CBCs—for example, friends introduced negative social norms in q49 and family back-

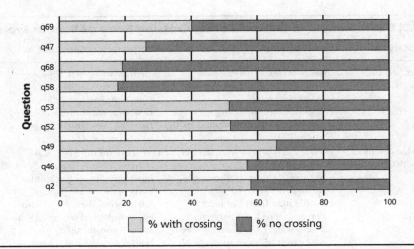

q2: Number of parents not educated or employed within science or technology (e.g., engineer, laboratory technician, computer scientist, scientist, veterinarian)

q46: "My parents' way of thinking is very different from the thinking of physics"

q49: "My best friends think that physics is boring"

q52: "You have to be a little odd/peculiar to work within physics"

q53: "My way of thinking is very different from the thinking of physics"

q58: "My parents take an interest in what goes on in school"

q68: "My best friends are serious about school work"

q47: "What goes on in physics class negatively affects my experiences with Nature by putting them into formulas and mathematics"

q69: "I consider it a problem that there is so little room for emotions in physics"

Figure 2.1   Frequency of exploratory Cultural Border Crossings.

grounds imposed distance from S&T (q2 and q46). On the other hand, some 80% of the students perceived that their parents and friends took an interest in schools and school work. This would indicate norms and support for schooling in general, and manageable transitions for most students into schools. Physics, however, seemed encumbered by frequent (>50%) value-oriented CBCs related to thinking style and negative images of science and scientists (q49, q52, q53). Less frequent, but still substantial, were affective border crossings related to students' perceiving physics/school physics as against their feelings. Since the study was only exploratory, no general conclusion about the relative frequencies of various types of border crossings could be extracted.

Establishing Scales and Exploring the Structure of CBCs

Since our intention was to add cultural perspectives to conventional attitudinal research (research question Q1), we had to establish a useful scale for attitudes towards science/physics that meets the conventional criteria of internal consistency and unidimensionality. This was accomplished with a three-component scale reflecting the multifaceted nature of the attitude construct:

Conative component: *"What is your inclination towards school physics"*
Affective component: *"Physics class is exciting"*
Cognitive component: *"In physics class we are working with things which are relevant and useful for life out of school"*

The items were found to be unidimensional with a Cronbach alpha of 0.76 and item-to-item correlations ranging from 0.41 to 0.67, well beyond the minimum acceptance level of 0.30 suggested by Gogolin & Swartz (1992).

Similarly, the structure of CBCs was studied by factor analysis (see Table 2.3). Within the above pool of nine CBCs, an interpretable structure

TABLE 2.3   Factor Loadings of Exploratory Cultural Border Crossings

	Factor			
Single item-CBCs	**Reputation**	**Home & Physics**	**Feeling**	**Home/ Peers & School**
q2: "Number of parents *not* educated or employed within Science or Technology."	−0.090	**0.754**	−0.162	0.006
q46: "My parents way of thinking is very different from the thinking of physics"	0.201	**0.777**	0.094	−0.110
q49: "My best friends think that physics is boring"	**0.761**	0.003	−0.066	−0.118
q52: "You have to be a little odd/ peculiar to work within physics"	**0.517**	−0.058	0.389	−0.040
q53: "My way of thinking is very different from the thinking of physics"	**0.620**	0.394	0.225	0.193
q53: "My way of thinking is very different from the thinking of physics"	0.236	−0.173	−0.246	**0.720**
q68: "My best friends are serious about school work"	−0.293	0.055	0.168	**0.752**
q47: "What goes on in physics class negatively affects my experiences with Nature by putting them into formulas and mathematics"	0.084	−0.083	**0.624**	−0.239
q69: "I consider it a problem that there is so little room for emotions in physics"	0.008	0.046	**0.762**	0.182

with four factors was found (Varimax Rotation, Kaiser-Guttman criteria, significant factor loadings > 0.3).

Most importantly, we found a factor structure consistent with the typology of CBCs: *Home & Physics* is a factor representing family background of relevance to physics, which is an example of a *Tradition-bound* CBC. The *Feeling* factor holds *affective* CBCs, while *Reputation* contains *Value-oriented* CBCs. Finally, the factor *Home/Peers & School* seems more of a *Value-oriented type CBC* than a *Tradition-bound* type.

Exploring the Impact of Various CBCs on Students' Attitudes and Subject Choices—Comparing with Other Constructs

Perceived cultural differences tend to complicate students' enculturation to science. However, as emphasized by Costa and others, border crossings can be smooth, manageable, hazardous, or even impossible, depending on the significance ascribed to each border crossing and the ways students' negotiate its meanings. So, more critical than the relative frequencies of various types of border crossings is their relative *significance*. Significance should be reflected in students' attitudes towards science—with obligatory caution that the causality of the relation cannot easily be established.

A number of conventional statistical techniques and procedures are available for the study of this relation. Figure 2.2 shows results from a Multiple Linear Regression analysis with attitude towards physics as dependent variable and a total of 17 constructs, including CBC factors, as explanatory variables (Krogh & Thomsen, 2005). Among the other constructs were personal variables (gender, school self-concept, physics self-concept), classroom variables (e.g., scales on student-centeredness and subject-centeredness), and teacher interactivity (e.g., enthusiasm, taking interest in students as persons).

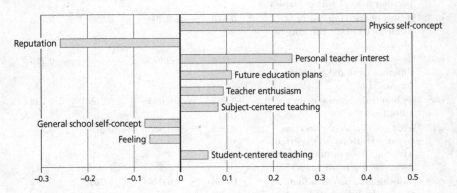

Figure 2.2 Standardized beta-coefficients from multiple regression analysis for attitude towards physics.

The inclusion of this larger set of constructs allows us to judge the importance of CBCs relative to more traditional measures. Here lies vital feedback to cultural studies, where the small sample sizes and thick descriptions tend to obscure more general patterns and proportions.

All standardized regression weights are significant at the 0.05 level or better. The predictors explain a total of 56% of the variation.

A much simpler model of comparable explanatory power (51%) is obtained when only three major explanatory variables are kept: *Physics self-concept,* value-oriented CBCs in *Reputation,* and the teacher interactivity component of *Personal teacher interest.* The analysis suggests that CBCs are among the most important influences on students' attitudes towards science, and particularly points towards value-oriented crossings related to the reputation and images of physics/school physics: Physics is perceived as strange, different, and boring.

Students in the sample had just chosen A-level subjects, and through Multiple Logistic Regression, a similar model for students' choice of A-level physics was obtained. Again the above three factors were highly important, even though *Gender* and *Future Education Plans (within S&T)* had larger β-weights. Also, the affective border crossings in *Feelings* seemed to have comparable impact on students' choice of A-level physics.

A simple yet powerful visualization of the impact of CBCs on students' choice of optional A-level physics was obtained when plotting the fraction of students with a given number of CBCs that actually pursue A-level physics. As seen by Figure 2.3, this fraction dramatically goes from 31% to half when two or more CBCs are present.

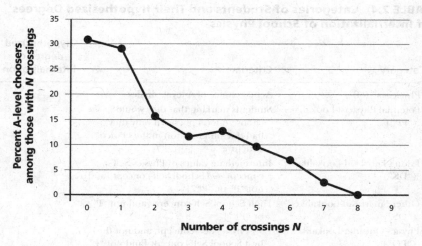

Figure 2.3   The relation between number of CBCs and probability of choosing A-level physics.

Again, causality cannot be established, but choice of A-level physics is clearly in *some way* related to the number of crossings. The issue, however, is complex since *Number of crossings* cannot be considered a unidimensional scale but rather an integration of four different factors.

Further Perspectives—CBCs and Internalization

"Internalization refers to the process whereby a person's affect toward an object passes from a general awareness level to a point where the affect is 'internalized' and consistently guides or controls the person's behavior" (Seels & Glasgow, 1990, p. 28). Usually, internalization is described by Kratwohl's "affective taxonomy" as proceeding through levels of *Receiving, Responding, Valuing, Organization,* and *Characterization by value or value set.* In our case, we had a variety of indicators somehow corresponding to different degrees of internalization: CBCs with and without values, students with many or few CBCs, students with or without physics/S&T-related future education plans, and students' actual choices in or out of S&T A-level subjects (physics, other science subjects, other subjects). Furthermore, students were grouped using a simple scheme (Table 2.4, criteria in second column).

The inspiration from Costa's (1995) student types is evident from the names. Costa's potential scientists were characterized by a profound orientation towards S&T education and careers. Students with actual (F) or almost-actual (PP) choice of A-level physics should be very similar to Costa's

TABLE 2.4 Categories of Students and Their Hypothesized Degrees of Internalization of School Physics

Category	Criteria	Hypothesized degree of internalization
"Physicists" ("P")	Actual choice of A-level physics	
"Potential Physicist-Lookalikes" ("PP")	Students marking that they would choose A-level physics if only they had one other option in the choice structure	
"I don't know kid-Lookalikes" ("DK")	Intermediate values of Physics Self-concept *and* School Self-concept [and not "P" or "PP"]	
"Other Smart Kid-Lookalikes" ("OS")	High School Self-concept [and not "P" or "PP"]	
"Physics Outsider-Lookalikes" ("PO")	Low Physics Self-concept, and not the best School Self-concept [and not "P" or "PP"]	

potential scientists in this aspect. Costa's other student types are classified according to their congruence with the world of school and the world of school science, respectively. Our classification in terms of general school self-concept and physics self-concept is at first sight very different. However, it should be noted that self-concept was the most important explanatory variable in the regression for attitudes towards physics. In that sense, self-concepts confound cognitive *and* attitudinal elements, and may capture important aspects of internalization—and of the Costa types. This last assumption was further supported when more elaborated group profiles were produced from our study (including all other aspects of our survey) and compared with Costa's thick descriptions.

Table 2.4 shows our initial hypotheses regarding student types representing different degrees of internalization. Our empirical setup allowed us to inquire into two important aspects of these hypotheses: first, from theoretical considerations we could not expect the variety of manifestations to be represented upon a unidimensional scale, but would empirical investigation substantiate a conception of internalization as *one* progression? Second, if so, is the relative order and progression caught correctly by our hypotheses? Correspondence Analysis was used to settle these questions, and the results are shown in Figure 2.4. Here, Plans for Future Education (PFE, with categories "Sc"—PFE within Science; "Te"—PFE within Technology; "No"—PFE outside S & T) and the student categories were

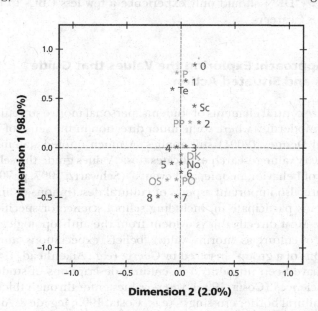

Figure 2.4 Correspondence analysis for internalization of (school) physics.

used as variables, and the number of border crossings was treated as supplementary variable.

With 98% of the variation located along Dimension 1, the analysis suggests that all of these variables may be used to describe aspects of the same progression. We see how indicators of high internalization consistently are placed at the top of the figure—for example, choice of A-level physics (P) or Future Education Plans within Science (Sc) or Technology (Te). Also we find the hypothesized sequence of student types, with the qualification that "Other Smart Kid-Lookalikes" ("OS") seem to be as far from internalizing the culture of physics/science as the "Outsider-Lookalikes" ("PO") – a qualification very much in accordance with Costa. Finally, we see the number of border crossing as an increasing sequence of (non-equidistant) markers along the "internalization axis." This suggests that internalization may be monitored, and maybe described, in terms of carefully crafted CBC-items. The figure also raises immediate questions for qualitative study, such as, How is it that actual Physics persisters are more oriented towards technological studies than scientific? Is low technology orientation the reason Potential Physicist-Lookalikes ended up choosing another A-level subject (typically from one of the "softer" science subjects)? For politicians and educators, the figure clearly shows that the group of "Other-Smart-Kid-Lookalikes" ("OS") is not a realistic target for pipeline-to-science-intervention. More comforting is the realization that the large group of "I Don't Know Lookalikes" ("DK") should only experience a few less CBCs to be within reach for S&T careers.

Second Approach: Exploring the Values that Guide Attitudes and Situated Action

Values are central elements of students' personal motive structures in the age of late modernity, where some inner direction in the sense of Riesman, Glazer, and Denney (2001) still applies. A much quoted statement from contemporary value research explicates this: "Values guide the selection or evaluation of behavior, people, and events" (Schwartz, 1997, p. 70).

Values are also important aspects of cultural descriptions of any subcultures students participate in, including school science or specific science classrooms. Most directly this is evident from the anthropological conceptualization of culture as "norms, values, beliefs, expectations, and conventional actions of a group" inspired by Geertz (e.g., Aikenhead, 1996).

Values have been introduced in cultural descriptions of students' "response to science" (Costa, 1995), where values enter through thick descriptions of "cultural border crossings" (e.g., Costa, 1995; Jegede & Aikenhead, 1999) or of students' construction of "science identities" (Carlone, 2004;

Brickhouse, Lowery, & Schultz, 2000). Typically, value aspects are only implicitly addressed in these studies. Cobern (1996) is an exception, as he explicitly addresses certain epistemological values in accordance with Kearney's Worldview Theory.

In the previous paragraph we saw how such cultural perspectives could be combined with a quantitative methodology, and how Cultural Border Crossings (CBCs) of value type were among the most important predictors of students' attitudes towards science and optional choice of A-level physics [ref]. While demonstrating the importance of such crossings compared to other factors, the repertoire of CBCs in this exploratory study was in no way complete. Now in this section, the aim is to build a more coherent cultural description of students' responses to science from the following line of thought: students participate in multiple worlds (family, friends, after-school activities, etc.) and bring their late modern identities and values from these worlds into the science classroom. Here they experience the subculture of school science and perceive cultural borders, some of them value-oriented. Guided by their personal value structures, students ascribe meanings/evaluations to the cultural differences (CBCs) and negotiate or otherwise select actions/responses to the Cultural Borders. The centrality of values lies in the fact that they are inherent in both structural and process-aspects of Cultural Border Crossing: perceived value differences (student–school science) are frequent and significant; students' values guide meaning-making, evaluation and actions. This is the fundamental argument driving the second approach described here. Both parts of the argument call for more systematic and complete descriptions of students' value systems and the values of science classrooms.

Other recent studies approaching "students' response" to science have used the term "value," but in a less strict sense than it will be used here. Thus, students' views on the "value of the science curriculum" have been exposed (Osborne & Collins, 2001), while their content preferences have been systematically surveyed by the ROSE project (Sjøberg & Schreiner, 2005) as "Values, interests, and priorities." In the context of the present approach, such content preferences or priorities would more properly be described as possible *indicators* of students' value orientations.

Values research has been conducted at least since Allport in 1931 (Allport, Vernon, & Lindzey, 1931), and significant bodies of contemporary research on values can be found in psychology, social psychology, sociology, ethnography, and market research. Even though some of these disciplines hold different perspectives on values (e.g., as individual respectively cultural constructs), a common core of understandings can be derived. As demonstrated below, science education may benefit from conceptual clarification and research instruments established within other disciplines.

A full program and a thorough unfolding of the potential of a value approach to the study of students' response to science would comprise at least the following points:

a. Conceptual clarification of the value construct and its relation to, for example, attitudes
b. Procedures and instruments to systematically investigate of students' value orientations
c. Procedures and instruments to systematically study the values of science classrooms
d. Alignment of the descriptions from (a) and (b), so they can be compared—in order to arrive at a systematic and more complete mapping of value-oriented CBCs than afforded by the previous exploratory study.
e. Exploration of students' value structures and how they guide students' situated evaluations and actions.

The full program has been elaborated in a PhD thesis (Krogh, 2006), but due to space limitations only points (a), (b), and (e) will be addressed here. These are most indispensable when it comes to indicating the potential of the methodology. However, a particular instrument for empirical analysis of values of science classrooms (TESSA: "The Ethos of School Science Analysis) has been devised (point c). TESSA maps the values of the science classroom along eight bipolar dimensions. A number of indicators for each dimension are specified, allowing empirical data of any kind to be used (curricula, survey, group interviews, textbooks, etc.). TESSA was invented through a hermeneutical process, where one consideration was to align its dimensions with important student value orientations (point d above). The use of TESSA to empirical characterization of the ethos of school physics has been reported in (Krogh, 2007), and slightly re-operationalized it has been used for textbook analysis.

Conceptual Clarification of the Value Construct and its Relation to Attitudes and Behavior

Some authors have seen a lack of consensus in the conceptual understanding of values in research studies (e.g., Van Deth & Scarbrough, 1995). However, this seems to be more of an issue with practitioners of research than a shortcoming within the existing theoretical basis. Readings of three seminal contributions to the foundations of value research (Kluckhohn, 1951; Rokeach, 1973; Schwartz, 1992) from cultural, psychological, and social-psychological perspectives indicate a reassuring consensus around the following defining characteristics of values:

- *Values are beliefs/conceptions that refer to desirable goals and to the modes of conduct that promote these goals* (Schwartz, 1997 p. 70). Similar notions are expressed in Rokeach (p. 6–9) and Klughohn (p. 395). Klughohn adds important comments, stating that values may be explicit or implicit and that they may characterize individuals as well as groups.

  Rokeach distinguishes between two types of values, *instrumental* ("desirable goals" or "end-states") and *terminal* ("modes of conduct"). Empirical work (cited in Schwartz, 1992, p. 49) however shows that changing items from one form (e.g. "politeness") to another (e.g. "being polite") is not reflected in any change of responses, which renders this distinction useless.

- *A finite set of values may approximately span the complete and (almost) universal value content.* Rokeach in his *Rokeach Value Survey* arrives at 36 values – from "intuitive, theoretical and empirical" considerations. Schwartz extends it to 56 values to have a "complete set" of "motivational types" represented in his semi-empirical *Schwartz's Value Survey* (SVS). Cross-cultural studies with SVS have demonstrated that the structure of closely related respectively opposing values ("value structure") is surprisingly stable across cultures. Schwartz himself uses the denotation "universal" (with "" markers), since "we do not believe that any single value structure is likely to be truly universal, so one must not generalize indiscriminately to new samples" (p. 47).

- *Values are internalized culture.* Schwartz (Schwartz, 1997, p. 70) states that "much of our mental programming is shared with other people ... This shared programming is what most social anthropologists and cross-cultural psychologists refer to as culture ... the heart of culture is formed by values—what people believe is good or bad, what they think should and should not be done, what they hold to be desirable or undesirable." As another part of our "mental programming," Schwartz emphasizes "unique sets of experiences and inherited traits." Similarly, Rokeach says, "The antecedents of human values can be traced to culture, society and its institutions, and personality" (p. 3). Values are social and cultural by nature, and get internalized in the meeting between individual biology, biography, and the social environment.

  Schwartz distinguishes theoretically and methodologically between personal and cultural values. He too provides empirical warranting behind his claim that value items can be phrased in such

a way that responses will be valid measures of personal values more
than cultural ideals (Schwartz, 1992).

- *Values are integrated into a relatively stable value structure.* Rokeach has
made an extensive test–retest analysis in relation to versions of RVS
with different ranking procedures for the values. With the optimal
ranking procedure (D), the test-retest reliability for a three-week
period was found to be 0.72–0.78. Over 14–16 months, this reliability dropped moderately to 0.61–0.69 for college students. Rokeach
finds these last estimates "especially noteworthy because they suggest an impressive degree of value system stability among college
students over relatively long periods of time" (Rokeach, 1973,
pp. 32–33). With younger students from grade 7, the stability is less
(reliabilities 0.53–0.62). The relative stability makes Rokeach speak
of values as "enduring beliefs." Schwartz has elaborated upon this
point: "Insofar as the basic human condition in which values are
grounded remains fairly constant, however, we anticipate that major
variations in value structure will be rare" (Schwartz, 1992, p. 47).

  Values are integrated in a relative value system, where *hierarchy*
describes the relative importance of values and *structure* describes the
opposition and relatedness of values. The value structure tends to be
the more stable, while value priorities (relative positions in the value-
hierarchy) may be influenced by situational activation: "After a value
is learned it becomes integrated somehow into an organized system
of values wherein each value is ordered in priority with respect to
other values. Such a relative conception of values enables us to define change as a reordering of priorities and, at the same time, to see
the total value system as relatively stable over time" (Rokeach, 1973).

- *"Values transcend specific actions and situations"* (Schwartz, 1997,
p. 70). Almost exactly the same formulation is used by Rokeach
(1973). This is a fundamental distinction between values and attitudes or interests. As emphasized by Fishbein and Ajzen (1975) and
Krapp and colleagues (Krapp et al., 1992), respectively, these latter
constructs describe a person-object relation. Values are more general constructs. Metaphorically the value system may be portrayed
as a personal gyroscope that tends to maintain a consistent personal
orientation across situations.

- *Values are motivational constructs that "guide the selection or evaluation
of behavior, people, and events"* (Schwartz, 1997, p. 71). Similar statements are found in Kluckhohn (1951, p. 395): "[values] influence
the selection of available modes, means and ends of action," and in
Rokeach: "Think of values as standards that guide ongoing activities, and of value systems as general plans employed to resolve
conflicts and to make decisions" (1973, p. 12).

An important point here is that values act as filters or structuring structures for judgments, attitudes, and actions. Along with this line of thinking, Rokeach elaborates on the centrality of values to self (1973, p. 18): "Values occupy a more central position than attitudes within one's personality makeup and cognitive system, and they are therefore determinants of attitudes as well as of behaviour."

Conceptually, this places values as precursors to attitudes and behavior, and interacting with both. While attitudes and behaviour are object-oriented and situated, students' value systems tend to be pretty stable across actions and situations for relatively long periods of time. These aspects have been indicated by adding values and relational arrows to one of the few models of attitudinal research, the Theory of Planned Behavior (Ajzen, 1991)—see Figure 2.5.

From theoretical considerations, one would also expect values to be co-determinants of students' Perceived Behavioral Control. Further, to account for (slow) changes/development of the value system one should add feedback arrows from TPB constructs to the personal value system, most plausibly from Behavior, Perceived Behavioral Control, and from Subjective Norms. Finally, to provide a model of *situated*, value-guided action, one needs mechanisms and processes describing how aspects of the situation interact with the general value system to produce outcomes of reasonable predictability, while still leaving room for students' agency. Metaphorically, one must establish the value gyroscope as a first step, and then look for aspects of the situation that might "flip" the gyro, temporarily changing saliency and hierarchy

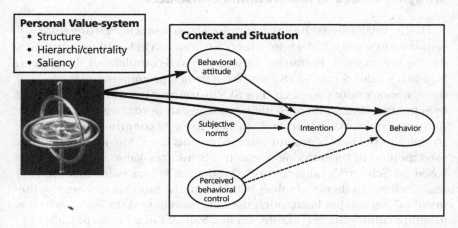

Figure 2.5 Values as general motivational structures behind attitudes, intentions, and situated action.

of central values in the structure. This would imply some arrows going back from Context and Situation to Personal Value System.

There is a considerable amount of empirical studies on how values are related to attitudes and behavior. However, they all come from other research fields than science education. Typically these studies do not apply an explicit social-psychological model, but use a more implicit argument that emphasizes values as "motivational constructs." More specifically, the importance of values has been demonstrated in relation to a span of different types of behavior, such as manifest voting behavior (Verplanken & Holland, 2002), pro-environmental behavior (Karp, 1996), consumer behavior (Grunert & Juhl, 1995), alcohol abuse (Dollinger & Kobayashi, 2003), and mothers' styles of conversation with their children (Tulviste & Kants, 2001). Hypothesized relations between certain values and types of behavior are generally significant, with typical correlations 0.2–0.4.

Similarly, there is a span of studies relating values to attitudes and personal interests. From a science educational point of view, some of the more interesting studies focus on vocational interest (Sagiv, 2002) and students' (intended) choices of various subjects (Feather, 1995). The studies largely confirm the hypothesized relations between certain values and preferences for future careers and subjects. This is concluded by Sagiv (e.g., 2002, p. 253), and correspondingly, Feather found that 80% of the hypothesized relations were significant with correlations 0.2–0.4. This level of indicators of the strength of the relation between values and preferences is found in most of the research literature.

Studying Values as Motivational Constructs

Having established values as motivational constructs underlying students' responses to science and school science we now turn to how these constructs can be investigated. Fortunately, there is a well-consolidated instrument, Schwartz's Value Survey (SVS), which has gained prominence in the field of contemporary values research. The SVS instrument was introduced in 1992 by Schwartz (1992), who presented it with an extensive cross-cultural validation on samples of students and teachers from 20 countries (N est. 8500). Its status in the field is evident from statements like: "The most commonly used method in recent value research is Schwartz's Value Survey, which is based on Schwartz's value theory" (Lindeman & Verkasalo, 2005, p. 157) and: "Schwartz's theory of values is at present the most comprehensive theory of values, and has been widely used in cross-cultural studies . . . as well as in within-culture studies" (Oishi, Diener, Suh, & Lucas, 1999, p. 162).

The SVS instrument was developed semi-empirically from a theory of motivational types. It has 56 items, each representing a general value. Schwartz argues that this list "encompasses virtually all the types of values to which individuals attribute at least moderate importance" (Schwartz, 1992, p. 59). Each item is construed as a general value in terminal or instrumental form, followed by an explanatory phrase in parentheses (e.g., the terminal item form "*Influential (having an impact on people and events*"). Respondents are supposed to rate the importance "as a guiding principle in my life" on a 9-point Likert-scale, ranging from *of supreme importance* (7) over *not important* (0) to *opposed to my values* (−1).

While theory originally informed the semi-empirical design of SVS, subsequent studies with the SVS in a reciprocal relationship have been used to consolidate Schwartz's Value Theory. In this way, Schwartz's value theory has established ten distinct motivational types, each characterized by a certain *content* of specific values. The value *content* of each motivational type has been tested and fine-tuned through *Smallest Space Analysis (SSA)* (Schwartz, 1992). Value *structure* here refers to the dynamic relations among the 10 value-constructs, stating their relative positions of, for example, opposition or congruence in value-space. The ten cross-national motivational types, their empirical content, and "almost universal" structure are described by Schwartz (1992). Here, the cross-cultural validation demonstrated that 87% of the values could be consistently assigned to a single motivational type in most of the samples (Schwartz, 1992). While supporting "almost universals" in the content and structure a few studies indicate (sub)cultural variation, a Portuguese study by Menezes and Campos (1997) exemplifies that, having 12 of the 56 values for a sample of students in upper secondary school fall into "wrong" facets. Most interestingly, samples of university students and younger teachers in the same study were assigned to the expected facets, indicating that the results for upper secondary students might be characteristic of the youth mentality/subculture, rather than national characteristics.

Schwartz's original approach relies on a specific multidimensional scaling (SSA) to position values within motivational types. However, factor analysis has also been used, for example, by Feather in a study of Australian university students' values (Feather, 1995). In this study the value structure was broken down to 14 factors, and a more detailed analysis revealed both subdivisions of motivational types and a few "rather different clusters." It remains unclear whether these differences should be attributed to the particular sample or to methodological features, since both analytical procedures introduce some choice and judgment (e.g., Schwartz, 1992, p. 45). The nature of interpretive flexibility differs from the semi-empirical SSA to the statistical factor analysis, but one method can hardly be warranted over the other.

The essential points here are that the SVS instrument applies across cultures, and that the exact number and content of motivational types shows minor variation with culture and sample. Western, late modern youth may be a particular case here. Finally and most importantly, the *distribution* of a youth sample across motivational types may vary much across context. The "hierarchy" question of how important various motivational types are within a (sub)cultural sample has not yet been addressed. For the study of students' response to science, this is the crucial question.

Analytical Steps: A Danish Example

A Danish study was set up to inquire into the general values of a sample of urban adolescents in upper secondary school, purposively sampled to capture youth mentalities shaped by Late Modernity. The sample consisted of 370 (200 female, 170 male) students from 17 classes in five general upper secondary *gymnasiums*, all located within Aarhus, the second largest city of Denmark with a population of approximately 300,000. There was an almost equal distribution of students in each of the three years of secondary schooling, corresponding to ages 17 to 19. Judged by all indicators of socio-cultural background (e.g., parents' educational backgrounds, index for number of books at home), the sample is dominated by adolescents from urban and privileged middle- or upper-class backgrounds.

Data were collected from a survey administered on an internet platform. A Danish translation of SVS was the core of the survey,[3] which also had 20 items inspired by the "*My Future Job*" part of the ROSE questionnaire (Sjøberg et al., 2005). These were added to check the relation between general values and science-related preferences.

First analytical step: Important values

The first step in the present analysis has been to identify single values of very high and remarkably low average importance as "guiding principles" in urban students' lives. Sample averages were calculated for girls and boys separately and compared by ANOVA to illuminate significant gender differences in the value structure. Table 2.5 reports the top 12 values in a gender-split format.

First of all, the table demonstrates that boys and girls are very similar in their ranking of values: Four of the top five values are identical, and nine of the top 12 values are the same. The impressive similarity suggests that gender-split analysis is only relevant when specifically addressing gender issues. Minor gender differences in value priorities can be found, and they produce a consistent pattern, since girls tend to give higher priority to *Honesty, Love,* and *Sense of Belonging*, while boys value *Pleasure, Intelligent,* and *Successful* more.

The many significant ANOVA measures are a little misleading here, since they express another (almost) consistent pattern: Girls generally seem to

TABLE 2.5   The Highest Ranked Values among Boys and Girls

Girls' Top 12 values	Average	Delta $(♀ - ♂)$	Boys' Top 12 values	Average	Delta $(♀ - ♂)$
1. True Friendship	5.68	0.69****	1. Freedom	5.10	0.05
2. Security (family)	5.49	0.98****	2. True Friendship	4.99	0.69****
3. Loyal	5.28	0.46***	3. Healthy	4.83	0.28*
4. Freedom	5.15	0.05	4. Loyal	4.81	0.46***
5. Healthy	5.12	0.28*	5. Pleasure	4.53	−0.36**
6. Responsible	5.11	0.72****	6. Security (family)	4.51	0.98****
7. Honest	4.97	0.75****	7. An Exciting Life	4.46	0.23
8. Love (mature)	4.95	1.11****	8. Successful	4.43	0.13
9. Sense of Belonging	4.80	0.60***	9. Intelligent	4.41	−0.60***
10. Choosing Own Goals	4.79	0.41***	10. Responsible	4.39	0.72****
11. Self-Respect	4.73	0.38**	11. Choosing Own Goals	4.38	0.41***
12. An Exciting Life	4.70	0.23	12. Self-Respect	4.35	0.38**

Note: Significance level of Delta $(♀ - ♂)$: * $p < 0.05$; ** $p < 0.01$; *** $p < 0.001$; **** $p < 0.0001$

attach higher importance to values as guiding principles in their personal lives than do boys. Significant gender differences ($p<0.05$) are found for 29 values, and in 26 of these cases the girls admit greater importance. So either girls are more into values or they are simply more willing to *admit* their value orientation.

The table and population averages indicate that positive students' responses would be guided most importantly by the following groups of values:

- Values of extremely high priority (average > 5 (out of 6)): *True Friendship, Freedom, Loyal* and *Security*
- Values of significantly high priority (average between 4.5 and 5.0): *Healthy, Responsible, Honesty, Choosing Own Goals, An Exciting Life, Self-Respect, Sense of Belonging, Successful*

And since deliberate *downgrading* of specific values may also act as a filter and a negatively reacting force some notion should be given to:

- Values of significantly low priority (average between 2.0 and 2.5): *An Effortless Life, Wealthy, Accepting My Portion in Life, Authority, Moderate, Respect for Tradition*
- Values of extremely low priority (average < 2): *Social Power, Devout, Unity with Nature*

Second Analytical Step: The Dominant Value Orientations

To distinguish the empirical value patterns of this study from Schwartz's ten semi-empirical motivational types, I will be using the term *value orien-*

tation. In the Danish study, value orientations were unidimensional, motivational constructs derived from factor analysis. Using the SAS procedure *FACTOR* with Ortogonal Varimax rotation and Kaiser-Gutman-criteria (Preacher & MacCallum, 2003, p. 21), this produced a total of 15 factors. The value content of each factor was derived from the general principle that values with a factor loading greater than 0.4 should be included. This is consistent with Preacher and MacCallum (2003, p. 27). All the prioritized guiding values from above are located within nine factors, separating all the high-priority values in five factors and the markedly low prioritized values within the remaining four. The detailed content, interpretation, and Cronbach Alpha's for the 5 essential and positive orientations are given in Table 2.6.

Here the right column of the table indicates the distribution of importance attached to the value orientations. As Cronbach alpha's of size 0.7 is generally accepted in the literature (Gogolin & Swartz, 1992; Streiner, 2003), the value orientations have reasonable quality.

Among the five high-prioritized value orientations, some 90% or more of the students consider these as "Highly important" or "Medium important" guiding principles in their personal lives! Almost the same fraction more

TABLE 2.6   The 5 High-Priority ("Dominant") Value Orientations and Their Priority Distribution Among Students

Factor interpretation High Priority Value orientations	Value content (factor loadings > 0.4)	Reliability of scale (Cronbach alpha)	Frequency of High–Medium–Low-importance attachment (%)
Togetherness-Relations	True friendship, Loyal, Security, Honest, Responsible	0.73	75–24–1
Autonomy/Self-direction	Freedom, Self-respect, Choosing own goals, Harmony, Pleasure ("gratification of desires"), Healthy	0.67	61–38–1
Identity-building Relations	Acceptance, Social recognition, Sense of belonging	0.67	42–47–11
Potentiation	An exciting life ("stimulating experiences"), Daring, Enjoying life, A varied life	0.69	45–45–10
Knowledge-Performance	Successful, Ambitious, Capable, Obedient ("dutiful, meeting obligations"), Wisdom	0.69	38–55–7

High: individual standardized scale score ≥ 4.5
Medium: 2.5 ≥ individual standardized scale scores < 4.5
Low: individual standardized scale scores < 2.5

or less turns away from the 4 markedly Low-prioritized value-orientations. There is every reason to believe that these orientations will influence students' judgments and actions in general, as well as their more specific responses to (school) science.

Among the high five, two value orientations are about *Relations*. First, the most important pattern, *Togetherness relations*, contains non-instrumental values characteristic of pure relations. In contrast, the pattern *Identity-building relations* holds values with a more instrumentalist benefit of entering relations. The value orientation *Autonomy/Self-Direction* is a mixture of values related to self-direction (e.g., *Choosing own goals*), being well-balanced (e.g., *Harmony*), and being physically fit in a self-assuring way (*Healthy*). The term *Potentiation* was inspired from the German sociologist of youth T. Ziehe, who defines this specific youth search pattern as: "*Potentiation* is creating meaning by artificial means. The search is neither for intimacy nor security, but for intensity" (Ziehe, 1989, p. 20). Finally, *Knowledge-Performance* is an integration of values that would resemble the ideal school code to almost any teacher.

Together, these five value orientations describe the most important motivational drives for *typical* urban adolescents of late modernity. To understand the evaluations and actions of *individuals*, one would have to determine their individual profiles along these orientations.

Value Orientations and Their Relation to Studies of Student Motivation

The value orientations established empirically in this study coincide with contemporary motivational research, first of all with Self-Determination Theory (SDT) of Ryan and Deci (2000). The value orientations of *Relations* (Togetherness/Identity building), *Autonomy/Self-Determination*, and *Knowledge-Performance* are in high accordance with the three essential "inner needs" of SDT, *Relatedness, Autonomy*, and *Competence*. Only the *Knowledge-Performance* component comprises both mastery and performance aspects in a more explicit manner than does the SDT competence.

The value orientations all can be used to guide both situated and long-term actions, but the *Potentiation* orientation tends to be more situational than the others. Mitchell (1993) has made a distinction between instructional CATCH and HOLD elements that should be supplied to stimulate students' interest (in math). The value orientations would seem to provide for both catch and hold elements, with *Potentiation* as the predominant reason to include CATCH elements in the sense of Mitchell.

Further, Perry (1992) has reviewed knowledge about informal learning environments that motivate. She finds that such environments should provide meaningful social interaction, sense of self-determination and control, sense of autonomy, and three "catchy" elements: challenge, curiosity, and play. Again, we see how these recommendations for motivating learning environments are in perfect correspondence with students' typical value orientations.

Finally, in the first approach we found that students' attitudes towards physics to a large extent could be explained by three explanatory variables: *Physics Self-concept, Reputation* (boring, strange, different), and *Personal teacher interest*. There the conception of values was narrower than the (almost) complete description of Schwartz used in the second approach. It is easy to see how these crucial explanatory variables match (elements of) the value-orientations *Knowledge-Performance, Potentitation,* and *Relatedness*, respectively.

So much, to argue that the dominant value-orientations provide explanation for typical trends in students' response to science. What remains in this article is to indicate how the approach can be extended to the study of individual and/or situated evaluation and action

Third Analytical Step: Individual Value-Profiles and Situated Responses

The dominant value-orientations were so frequent with students that a typical student will hold several such value-patterns simultaneously. A closer inspection of the data reveals that a typical student will hold three of the High-priority and two of the Low-priority orientations at the same time. These span the axes of the value-gyroscopes students bring to the science classroom. For pedagogical action on the *class level* it may suffice for teachers to know the 5 important value-orientations and how to interact with them. However, understanding *individual* students' judgments and actions requires at least a fundamental knowledge of their personal profile of dominant value-orientations. Further one must have an idea of the relative hierarchy of these and of situational mechanisms that may activate or reinforce specific value-orientations. These aspects will be discussed below.

In an action research project undertaken by the author in 2005, 26 students within an upper secondary physics B class were studied. Figure 2.6 illustrates that on top of a background of shared student values there still is diversity and individual uniqueness.

The two-year-long action research project had as its primary aim to meet students' individual value orientations (as far as possible) through deliberate teaching design. At the same time, it provided opportunities to investigate how value orientations guide students' preferences and actions within the classroom. Both purposes required reliable measures of value orientations, which were obtained with the use of SVS and a member-checking process: Individual value profiles were calculated, and each value profile was converted into a verbal, written description. Descriptions were generated by inserting pre-written phrases according to each level of value orientation support, and adjusting the result for linguistic coherence. These verbal representations were fed back to all students, who were urged to comment: Did they approve of the value description of themselves? For more elaborate validation and triangulation, a group of 12 focus students

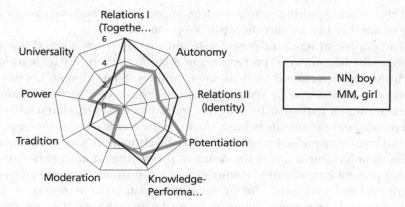

Figure 2.6 Sample value profiles of two students.

were interviewed using a modified AIMS interview protocol (Wadsworth & Ford, 1983), which integrated students' responses to their written SVS profiles. The students were almost surprisingly approving of the descriptions; only two students expressed some reservation about parts of their profiles, maintaining that some component was too extreme or that some element of the profile was only valid for certain contexts. Overall, the validation procedures gave reassurance to the interpretation of the digital SVS value profiles of individual students.

Analyzing the Relation Between Value Orientations and Students' Preferences in the Science Classroom Context

The claim from values research that value orientations guide students' evaluations and actions was tested in various ways. Only one will be detailed here.

Students' evaluations and preferences in relation to the ongoing physics class were systematically monitored with the use of a specially designed evaluation sheet ("Judging the Temperature of Physics," or JTP) after every larger thematic learning unit. Students were supposed to evaluate their perception of learning and being in the physics classroom along 25 bipolar dimensions (e.g., dimensions of social climate, aspects of the use of activities, teacher interactivity, topic relevance, student involvement in decision making, student autonomy and options, etc.). A number of the dimensions were aligned with significant value orientations. Each dimension was evaluated with two questions, one capturing *Perceived characteristics of the present state* (5-point Likert scale between the two poles of the specific dimension), and the other capturing *satisfaction with present state or preferred changes* in one or another bipolar direction (three options here).

The double structure of *present state* and *wished-for state* allowed one to analyze whether students' preferences in relation to their physics learning environment correspond to their value profiles. Some double combinations of responses may be seen as certain indicators of a particular value-orientation at a particular level of hierarchy. For example, characterizing the present state as already being a "*little oriented towards*" the *Variation* pole and still *wishing for more Variation* would positively identify with a *High orientation towards Potentiation*. On the contrary, *not* wishing for more variation in such a present state situation disqualifies as a consistent *High orientation towards Potentiation* response. Finally, some combinations of present state and wished-for-state responses fall into an inconclusive grey area. If for example the present state is characterized as a "*little oriented towards*" the *Monotony* pole (opposite *Variation*), then wishing for more *Variation* may express a *Moderate*, as well as a *High orientation* towards *Potentiation*. Analyzing the possible response matrices more systematically shows that five out of the 15 combinations may be taken as positive identifications, no matter if we are looking for a *High, Moderate,* or *Low* student orientation. A larger amount (4/9) of combinations represent incidents disconfirming value types, while the remaining number of combinations represent inconclusive cases. The important point is here that the actual distribution of value-congruent, value-discordant and grey-area incidents in the JTP analysis can be compared to the distribution in case of randomized double answers: 1/3, 4/9, and 2/9. Table 2.7 shows the results from analysis of 559 contextualized preferences captured on three different occasions during the research project.

In other words, students' instructional preferences in the context of school science do show considerable correspondence with their individual value profiles (Chi-square-statistics of 156.8, 2 degrees of freedom, $p < 1E-6$).

The methodology could easily be used to study students' *situated* responses within the school science context. Combining value profiles with situated JTP data and videotapes of instructional activities would make it possible to investigate matters of value system dynamics, first of all the processes of value-*saliency* and *activation*. Within the research-literature issues of saliency and hierarchy have been approached by Verplanken and Holland (2002), introducing the concept "Centrality to Self" and concluding: "Values were found to give meaning to, energize, and regulate value-congruent behavior, but only if values were cognitively activated and central to the self" (p. 434). Processes of activation through context factors have been explored by Higgins (1996), and Ford (1992) has illuminated how the possible *alignment* of several goals and values across a single "behaviour episode" enhances the tendency to act accordingly. Knowing students' value profiles ("value gyros") and how value

TABLE 2.7 Frequency of Value-Congruent and Value-Discordant Evaluations in the Context of School Physics

Value orientation	Corresponding Value poles of JTP	Value-congruent preferences	Value-discordant preferences	Inconclusive incidents
Relation I (Togetherness)	Friendship (dim2):	41	5	0
	Security (dim3):	44	7	0
Relation II (Identity)	Belonging (dim1):	35	2	6
	Recognition (dim4):	34	6	12
Autonomy/Self-direction	Following own goals (dim10a):	17	7	12
	Cooperation (dim5, inverted):	12	21	23
	Personal development (dim21)	35	8	11
Potentiation	Excitement (dim14):	29	8	19
	Variation (dim9):	30	10	16
Knowledge-Performance	Success (dim24):	17	21	15
	Achievement (dim22):	16	19	21
TOTAL		310	114	135
		55%	20%	24%
RANDOM (for comparison)		33%	44%	22%

systems may interact with contextual factors are the essential prerequisites to understanding students' situated responses to school science.

CONCLUDING REMARKS

The two approaches presented in this chapter extend conventional attitudinal research. They enter cultural perspectives, emphasizing cultural border crossing and values as critical influences, not only on students' attitudes towards science, but on students' more general responses to (school) science. Methodologically they apply quantitative measures *and* open for thick descriptions to produce understanding of students' responses. The second approach with values as motivational constructs may be considered a stand-alone methodology, or it may be considered a special case of the first. To explicate values as a special case of cultural border crossings, one needs to map the mismatches between students' personal value systems and the ethos of school sciences. Such a full approach has been applied in a systematic study of students' value-oriented cultural border crossings in relation to upper secondary school physics in Denmark (Krogh, 2007; Krogh, 2006). The second approach with its combination of personal value profiles and situational triggers also meets the challenge of balancing structural and dynamic aspects of students' responses to science, and it provides one possible framework to handle the problem of a missing ontogenetic perspective within attitudinal/interest research emphasized by Krapp (2002). Elaboration of this aspect would seem a valuable next contribution.

NOTES

1. One might just as well have used qualitative (e.g., interview) data, and some sort of coding scheme to produce categorical or nominal data.
2. $N = 789$, students in 2nd year of Danish Upper Secondary School (Gymnasium, stx).
3. By mistake the internet version of the questionnaire assigned the number "6" to the category "*of supreme importance*" instead of 7 as in the original SVS. While introducing a new maximum, there is no reason to believe that this in any way will distort the results

REFERENCES

Aikenhead, G. S. (1996). Science education: Border crossing into the subculture of science. *Studies in Science Education, 27,* 1–52.

Aikenhead, G. S. (2001). Students' ease in crossing cultural borders into school science. *Science Education, 85,* 180–188.

Ajzen, I. (1985). From intentions to actions: a theory of planned behavior. In J.Kuhl & J. Beckham (Eds.), *Action-control: From cognition to behavior* (pp. 11–39). New York: Springer-Verlag.

Ajzen, I. (1991). The theory of planned behavior. *Organizational Behavior and Human Decision Processes, 50,* 179–211.

Allport, G. W., Vernon, P. E., & Lindzey, G. (1931). *Study of values: A scale for measuring the dominant interests in personality.* Boston: Houghton Mifflin.

Blickenstaff, J. C. (2005). Women and science careers—Leaky pipeline or gender filter. *Gender and Education, 17,* 369–386.

Brickhouse, N. W. & Potter, J. T. (2001). Young women's identity formation in an urban context. *Journal of Research in Science Teaching, 38,* 965–980.

Brickhouse, N. W., Lowery, P., & Schultz, K. (2000). What kind of a girl does science? The construction of school science identities. *Journal of Research in Science Teaching, 37,* 441–458.

Brown, B. A. (2004). Discursive identity: Assimilation into the culture of science and its implications for minority students. *Journal of Research in Science Teaching, 41,* 810–834.

Carlone, H. (2005). *Science identity in science education: Possibilities and complexities.* http://conferences.uconn.edu/crossroads/download/HCarlone.pdf

Carlone, H. B. (2004). The cultural production of science in reform-based physics: Girls' access, participation, and resistance. *Journal of Research in Science Teaching, 41,* 392–414.

Chinn, P. W. U. (2002). Asian and Pacific islander women scientists and engineers: A narrative exploration of model minority, gender, and racial stereotypes. *Journal of Research in Science Teaching, 39,* 302–323.

Cobern, W. W. (1996). Worldview theory and conceptual change in science education. *Science Education, 80,* 579–610.

Costa, V. B. (1995). When science is another world—Relationships between worlds of family, friends, school, and science. *Science Education, 79,* 313–333.

Denzin, N. & Lincoln, Y. E. (2005). *Qualitative research.* (3rd. ed.). Thousand Oaks, CA: Sage.

Dollinger, S. J. & Kobayashi, R. (2003). Value correlates of collegiate alcohol abuse. *Psychological Reports, 93,* 848–850.

Feather, N. T. (1995). Values, valences, and choice: The influence of values on the perceived attractiveness and choice of alternatives. *Journal of personality and social psychology, 68,* 1135–1151.

Fishbein, M. & Ajzen, I. (1975). *Belief, attitude, intention, and behavior: An introduction to theory and research.* Reading: MA: Addison-Wesley.

Ford, M. E. (1992). *Motivating humans—Goals, emotions and personal agency beliefs.* London: Sage Publishers.

Gardner, P. L. (1975). Attitudes to science: A review. *Studies in Science Education, 2,* 1–41.

Gardner, P. L. (1995). Measuring attitudes to science: Unidimensionality and internal consistency revisited. *Research in Science Education, 25,* 283–289.

Geertz, C. (1973). *The interpretation of cultures.* New York: Basic Books.

Giddens, A. (1991). *Modernity and self-identity: Self and society in the late modern age.* Stanford, CA: Stanford University Press.

Gilbert, S. & Yerrick, R. (2001). Same school, separate worlds: A sociocultural study of identity resistance, and negotiation in a rural, lower track science classroom. *Journal of Research in Science Teaching, 38,* 574–598.

Gogolin, L. & Swartz, F. (1992). A quantitative and qualitative inquiry into the attitudes toward science of nonscience college students. *Journal of Research in Science Teaching, 29,* 487–504.

Grunert, S. C. & Juhl, H. J. (1995). Values, environmental attitudes, and buying of organic foods. *Journal of Economic Psychology, 16,* 39–62.

Guba, E. & Lincoln, Y. (2005). Paradigmatic controversies, contradictions, and emerging confluences. In N.Denzin & Y. Lincoln (Eds.), *Handbook of qualitative research* (3rd ed., pp. 191–215). Thousand Oaks, CA: Sage Publications.

Higgins, E. T. (1996). Knowledge activation: Accessibility, applicability and salience. In E.T.Higgins & A. W. Kruglanski (Eds.), *Social psychology: Handbook of basic principles* (pp. 133–168). New York: Guilford.

Husén, T. (1997). Research paradigms in education. In J.P. Keeves (Ed.), *Educational research, methodology, and measurement: An international handbook* (2nd ed., pp. 16–21). Oxford: Elsevier Science Ltd.

Jegede, O. & Aikenhead, G. (1999). Transcending cultural borders: Implications for science teaching. *Research in Science & Technological Education, 17,* 45–66.

Johnson, R. B. & Onwuegbuzie, A. J. (2004). Mixed methods research: A research paradigm whose time has come. *Educational Researcher, 33,* 14–26.

Karp, D. G. (1996). Values and their effect on pro-environmental behavior. *Environment and Behavior, 28,* 111–133.

Kluckhohn, C. K. (1951). Values and value orientations in the theory of action. In T. Parsons & E. A. Shils (Eds.), *Toward a general theory of action* (pp. 388–433). Cambridge, MA: Harvard University Press.

Krapp, A. (2002). Structural and dynamic aspects of interest development: theoretical considerations from an ontogenetic perspective. *Learning and Instruction, 12,* 383–409.

Krapp, A., Hidi, S., & Renninger, K. A. (1992). Interest, learning and development. In K.A.Renninger, S. Hidi, & A. Krapp (Eds.), *The role of interest in learning and development* (pp. 3–25). Hillsdale, NJ: Lawrence Erlbaum Associates.

Krogh, L. B. (2006). *"Cultural border crossings" within the physics classroom—A cultural perspective on youth attitudes towards physics.* Steno Department for Studies of Science and Science Education (In Danish).

Krogh, L. B. (2007, August). *The ethos of school science—An analytic framework and the empirical characterization of values and norms within school physics.* Paper presented at the European Science Educational Research Association (ESERA), Malmö, Schweden.

Krogh, L. B. & Thomsen, P. V. (2005). Studying students´ attitudes towards science from a cultural perspective but with a quantitative methodology: border crossing into the physics classroom. *International Journal of Science Education, 27,* 281-302.

Lindeman, M. & Verkasalo, M. (2005). Measuring values with the short Schwartz's value survey. *Journal of Personality Assessment, 85,* 170–178.

Menezes, I. & Campos, B. (1997). The process of value-meaning construction: A cross-sectional study. *European Journal of Social Psychology, 27,* 55–73.

Mitchell, M. (1993). Situational Interest: Its multifaceted structure in the secondary school mathematics classroom. *Journal of Educational Psychology, 85,* 424–436.

Oishi, S., Diener, E., Suh, E., & Lucas, R. (1999). Value as a moderator in subjective well-being. *Journal of Personality, 67,* 157–184.

Osborne, J. & Collins, S. (2001). Pupils' views of the role and value of the science curriculum: A focus-group study. *International Journal of Science Education, 23,* 441–467.

Osborne, J., Simon, S., & Collins, S. (2003). Attitudes towards science: A review of the literature and its implications. *International Journal of Science Education, 25,* 1049–1079.

Osborne, J., Simon, S., & Tytler, R. (2009, April). *Attitudes towards science—An update.* Paper presented at the Annual Meeting of the American Educational Research Association, San Diego, California. http://www.kcl.ac.uk/content/1/c6/05/84/75/AttitudesTowardScience.pdf .

Perry, D. (1992). What research says: Designing exhibits that motivate. *ASTC Newsletter, 20,* 9–12.

Phelan, P., Davidson, A. L., & Cao, H. T. (1991). Students multiple worlds—Negotiating the boundaries of family, peer, and school cultures. *Anthropology & Education Quarterly, 22,* 224–250.

Preacher, K. J. & MacCallum, R. C. (2003). Reparing Tom Swift's electric factor analysis machine. *Understanding Statistics, 2,* 13–43.

Ramsden, J. (1998). Mission impossible? Can anything be done about attitudes to science? *International Journal of Science Education, 20,* 125–137.

Riesman, D., Glazer, N., & Denney, R. (2001). *The lonely crowd.* New Haven, CT: Yale University Press.

Rokeach, M. (1973). *The nature of human values.* New York: The Free Press.

Ryan, R. M. & Deci, E. L. (2000). Self-determination theory and the facilitation of intrinsic motivation, social development, and well-being. *American Psychologist, 55,* 68–78.

Sagiv, L. (2002). Vocational interests and basic values. *Journal of Career Assessment, 10,* 233–257.

Schibeci, R. A. (1984). Attitudes to science: an update. *Studies in Science Education, 11,* 26–59.

Schreiner, C. (2006). *Exploring a ROSE-garden: Norwegian youth's orientations towards science—seen as signs of late modern identities.* Faculty of Education, University of Oslo.

Schwartz, S. (1992). Universals in the content and structure of values: Theoretical advances and empirical tests in 20 countries. In M.P.Zanna (Ed.), *Eksperimental social psychology* (pp. 1–65). San Diego: Academic Press.

Schwartz, S. (1997). Values and culture. In D.Munro, J. F. Schumaker, & S. C. Carr (Eds.), *Motivation and culture* (pp. 69–84). New York and London: Routledge.

Seels, B. & Glasgow, Z. (1990). *Exercises in instructional technology.* Columbus, OH: Merrill Publishing.

Sjøberg, S. & Schreiner, C. (2005). ROSE (Relevance Of Science Education). http://www.ils.uio.no/forskning/rose/.

Streiner, D. (2003). Starting at the beginning: An introduction to coefficient alpha and internal consistency. *Journal of Personality Assessment, 80,* 99–103.

Tulviste, T. & Kants, L. (2001). Conversational styles of mothers with different value priorities: Comparing Estonian mothers in Estonia and Sweden. *European Journal of Psychology of Education, 16,* 223–231.

Van Deth, J. W. & Scarbrough, E. (1995). The concept of values. In J.W. Van Deth & E. Scarbrough (Eds.), *The Impact of Values* (pp. 21–47). New York: Oxford University Press.

Verplanken, B. & Holland, R. W. (2002). Motivated decision making: Effects of activation and self-centrality of values on choices and behavior. *Journal of personality and social psychology, 82,* 434–447.

Wadsworth, M. & Ford, D. H. (1983). Assessment of personal goal hierarchies. *Journal of Counseling Psychology, 30,* 514–526.

Waldrip, B. G. & Taylor, P. C. (1999). Permeability of students' worldviews to their school views in a non-western developing country. *Journal of Research in Science Teaching, 36,* 289–303.

Walker, J. C. & Evers, C. W. (1997). Research in education: Epistemological issues. In J.P. Keeves (Ed.), *Educational research, methodology, and measurement: An international handbook* (pp. 40–56). Oxford: Elsevier Science Ltd.

Weber, M. (1947). *The theory of social and economic organization.* New York: The Free Press.

Zeuner, L. & Linde, P. C. (1997). *Life strategies and choices of education* Copenhagen: The Danish National Institute of Social Research. In Danish.

Ziehe, T. (1989). *Ambivalens og mangfoldighed [Ambivalence and Multiplicity].* København/Copenhagen: Politisk Revy.

Ziehe, T. (2004). *Islands of intensity on a sea of routine—New articles on youth, education, and culture.* Copenhagen: Politisk Revy (Anthology in Danish).

CHAPTER 3

DEVELOPMENT AND TEST OF AN INSTRUMENT THAT INVESTIGATES TEACHERS' BELIEFS, ATTITUDES AND INTENTIONS CONCERNING THE EDUCATIONAL USE OF SIMULATIONS

Zacharias C. Zacharia, Ioanna Rotsaka, and Tasos Hovardas
University of Cyprus

ABSTRACT

The purpose of this research was to develop an instrument, namely, a questionnaire that investigates teachers' beliefs, attitudes, and intentions concerning the use of simulations for educational purposes, including for teaching science. The development of such an instrument is needed because of the absence of an instrument that specifically targets teachers and simulation use for educational purposes. To date, the only questionnaires that are available in this domain focus on technology or computer technologies in general, which from

Attitude Research in Science Education, pages 81–115

our perspective are not proper for investigating teachers' beliefs, attitudes, and intentions towards the use of simulations, in the sense that we do not know how each teacher defines or understands the term computer technology. The development of the questionnaire was based on attitude–behaviour theories/ models developed in prior research studies focusing on the affective domain and computer technologies. For validity and reliability purposes, we collected data from 514 elementary school teachers who were aware of what a simulation is and taught science before this study. All participants were randomly·selected. Structural Equation Modeling (SEM) was also implemented on the data captured in order to reach to a possible simulation acceptance model. Finally, besides reporting on the development of this instrument and its underlying model, we present findings that concern our teachers' beliefs, attitudes, and intentions concerning the use of simulations for educational purposes.

INTRODUCTION

Although the concepts of beliefs, attitudes, and intentions concerning the use of computers have gained recognition as critical determinants in the adoption and acceptance of computer technologies (Agarwal & Prasad, 1999; Barki & Hartwick, 1994; Culpan, 1995; Davis, 1989; Davis, Bagozzi, & Warshaw, 1989; Koslowsky, Holman, & Lazar, 1990; Smarkola, 2007; Teo, Lee, & Chai, 2008; Teo, Su Luan, & Sing, 2008), there is no single, universally accepted definition of the computer construct (Kay, 1992; Zanna & Rempel, 1988; Zacharia, 2003). For instance, Kay (1993) noted that attitudes towards computers have been defined in at least 14 different ways in the computing research literature. Despite all the efforts to come to a consensus, this problem has still not been resolved (Smith, Caputi, & Rawstorne, 2000; Zacharia, 2003). One of the reasons behind this variability is the fact that the computer construct is very broad. In fact, it involves all the available computer-based technologies to date (e.g., multimedia, animations, simulations, games). Therefore, a clear definition of the computer construct is needed before developing and using any kind of attitudinal instrument in order to get valid measures about any computer technology.

One approach to overcome this problem is to identify the computer technology that we want to measure individuals' beliefs, attitudes, and intentions toward and then proceed with the development of an instrument that specifically targets that particular computer technology. Using the computer as a blackbox construct implies the belief that all people understand the same thing when they are confronted with the computer construct, which is not true. Any measure against to such a generalized backdrop could produce invalid, even deceiving in some ways, results. For instance, are students' attitudes towards computer games and text editors the same? For valid measures about individuals' beliefs, attitudes, and intentions about any computer-based technology construct, researchers need to define/specify both the construct and its context, as well as, to make this construct and its context apparent to the individuals that are undertaking the survey.

For the purposes of this study we developed and tested an instrument that investigates teachers' beliefs, attitudes, and intentions concerning the educational use of simulations. In this case, simulation is the computer-based technology construct, and teachers and the use of simulations for educational purposes comprise the context in which we situated this construct. We further aimed to investigate whether the use of simulations was associated with science teaching, rather than any other subject domain teaching. Another aspect of this study was to explore a model to understand teacher acceptance of simulation use. Such an instrument is particularly important because it could provide us with a picture as far as whether a teacher intends to use a proven efficient teaching and learning tool, such as a simulation (Triona & Klahr, 2003; van der Meij & de Jong, 2006; Zacharia & Anderson, 2003; Zacharia, 2007; Zacharia, Olympiou, & Papaevripidou, 2008), for educational purposes.

Prior studies have stated that teachers' beliefs, attitudes, and intentions as well as knowledge and skills in using computers influence their initial acceptance of computer technology and their future behavior regarding computer usage (Koohang, 1989; Lawton & Gerschner, 1982; Violato, Mariniz, & Hunter, 1989; Zacharia, 2003), as well as instructional use of computers and likelihood of profiting from training (Kluever, Lam, & Hoffman, 1994). However, recent studies find that teachers and students are still unwilling to engage in an active or sustained manner in activities using computer technologies (Becker, 2001; Breen, Lindsay, Jenkins, & Smith, 2001; Burns, 2002; Marriott, Marriott, & Selwyn, 2004; Reffell & Whitworth, 2002; Wozney, Venkatesh, & Abrami, 2006), including computer simulations. Becker (2001) found that teachers generally used computer technology to support their existing practices (providing practice drills, demonstration) and communication (such as the use of email) rather than to engage students in learning that involves higher order thinking.

On the other hand, other researchers present computer technology applications as an integral part of teaching and learning due to their direct impact on current educational practices and policies and their potential to transform education (Bereiter & Scardamalia, 2006; Zacharia & Constantinou, 2008; Zacharia et al., 2008). Therefore, it is important for teachers to understand the precise role of each computer technology so that they can effectively cope with the pressure created by continual innovation in educational technology and tensions to prioritize the use of computer technology. Of course, this creates the need to predict and understand teachers' computer technology use and acceptance due to their strong link to classroom actions. Overall, the importance of targeting teachers' beliefs, attitudes, and intentions to change their teaching behaviors has been recognized in many education contexts, including science education (Bencze, & Hodson, 1999; Tobin, Tippins, & Gallard, 1994; Zacharia, 2003). The purpose of this study was to contribute towards this direction and develop a Simulation Acceptance Model (SAM).

In developing an instrument that measures teachers' beliefs, attitudes, and intentions, we have turned to the Technology Acceptance Model (TAM) (Davis, 1989; Davis, Bagozzi, & Warshaw, 1989), which is grounded in the attitude–behavior Theory of Reasoned Action (TRA) (Ajzen & Fishbein, 1980; Fishbein & Ajzen, 1975; Fishbein, 1980). We have selected TAM because, over the years, researchers have successfully used the TAM framework to examine users' acceptance toward several computer-based technology constructs, such as the Graphic User Interface (GUI) (Agarwal & Prasad, 1999), mainframe application (Dishaw & Strong, 1999), accounting applications (Jackson, Chow, & Leitch, 1997), World Wide Web (Moon & Kim, 2001; Riemenschneider, Harrison, & Mykytn, 2003), e-learning (Yuen & Ma, 2008), and computer resource center (Taylor & Todd, 1995). Moreover, researchers have used the TAM to investigate education-related issues such as student satisfaction with online learning (Drennan, Kennedy, & Pisarksi, 2005), student acceptance of online course companion site of a textbook (Gao, 2005), and the effect of technical support on student acceptance towards WebCT (Ngai, Poon, & Chan, 2007). Nonetheless, there is no study that used the TAM framework to examine users' acceptance of simulations.

THEORETICAL FRAMEWORK

In the last three decades, researchers have given much attention to identifying the conditions or factors that could facilitate technology, including computer technology, integration into businesses (Legris, Ingham, & Collerette, 2003) and education (Ma, Andersson, & Streith, 2005; Teo, Lee, & Chai, 2008; Teo, Su Luan, & Sing, 2008). Arising from this motivation, models were developed and tested to help in predicting technology acceptance. Among these models, the Technology Acceptance Model (TAM) has been the most widely adopted theoretical framework to study technology acceptance (Ma et al., 2005; McCoy, Galletta, & King, 2007; Teo, Lee, & Chai, 2008; Teo, Su Luan, & Sing, 2008; Yuen & Ma, 2002, 2008). Introduced by Davis (1989), TAM was an adaptation of the Theory of Reasoned Action (Ajzen & Fishbein, 1980) specifically tailored for modeling how users come to accept and use technology. Over the years, Davis and colleagues (Davis, 1989; Davis, Bagozzi, & Warshaw, 1989; Venkatesh, 1999; Venkatesh & Davis, 1996, 2000; Venkatesh, Morris, Davis, & Davis, 2003) conducted numerous experiments to validate TAM. Overall, the TAM has been empirically proven successful in predicting about 40% of technology use (Legris et al., 2003). It has also found to be a parsimonious representation of how beliefs/perceptions and attitudes affect technology use (Sivo & Pan, 2005).

As shown in Figure 3.1, TAM links the *Perceived Ease of Use* (PEU) and *Perceived Usefulness* (PU) with attitude (AT) towards using computers, behavioral intention (BI), and actual use (system use). The model suggests that when users are presented with a computer technology, a number of variables influence their decisions about how and when they will use it. There are two specific variables, PEU and PU, which are hypothesized to be fundamental determinants of user acceptance (Davis et al., 1989). The PEU is "the degree to which the...user expects the target system to be free of effort" (Davis et al., 1989, p. 985). It is possible that while users may believe that computers are useful, they may be, at the same time, too difficult to use, and that the performance benefits of usage are outweighed by the effort of using the application (Davis, 1989). PEU explains the user's perception of the amount of effort required to utilize the system or the extent to which a user believes that using a particular technology will be effortless (Davis et al., 1989). PU is the user's "subjective probability that using a specific application system will increase his or her job performance within an organizational context" (p. 985). In the view of Phillips, Callantone, and Lee (1994), PU reflects the prospective users' subjective probability that applying the new technology will be beneficial to his/her personal and/or the adopting organization's well-being. As such, it is possible that educational technology with a high level of both PEU and PU is more likely to induce positive attitudes. According to Moon and Kim (2001), PU has direct impacts on attitude, while PEU influences attitude indirectly through PU. On the other hand, Yuen and Ma (2002) found that the PEU

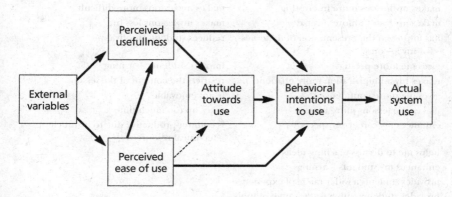

Figure 3.1 Technology acceptance model (Davis, Bagozzi and Warshaw, 1989).

and PU directly affect the intention to engage in computer use as stated in the TAM.

The *external variables*, shown in Figure 3.1, vary according to the context of the construct under study. In the case of our study, the *external variables* represent the many influences on teachers, which come from outside their sphere of control. These will include aspects, such as teacher's computer skills, school policies on using computers, opinions of colleagues, responsibilities of the teacher, pressure from parents and students, and the influence of the education authority. Research has identified a number of positive and negative factors influencing the PEU (e.g., Watson, 1993) and the PU (e.g., Askar & Umay, 2001; Cox, Preston, & Cox, 1999). Examples of these factors are given in Tables 3.1 and 3.2.

TABLE 3.1   Positive and Negative Factors Influencing PEU (Cox, Preston, & Cox, 1999)

Positive factors	Negative factors
regular use and experience of computers outside the classroom	difficulties in using software/hardware
ownership of a computer	need more technical support
confidence in using computers	not enough time to use computers
easy to control the class	is too expensive to use regularly
easy to think of new lesson ideas	insufficient access to the resources
can get help and advice from colleagues	restricts the content of the lessons

TABLE 3.2   Positive and Negative Factors Influencing PU (Cox, Preston, & Cox, 1999)

Positive factors	Negative factors
makes my lessons more interesting	makes my lessons more difficult
makes my lessons more diverse	makes my lessons less fun
has improved the presentation of materials for my lessons	reduces students' motivation
gives me more prestige	impairs students' learning
makes my administration more efficient	restricts the content of the lessons
gives me more confidence	is not enjoyable
makes the lessons more fun	takes up too much time
enhances my career prospects	is counter-productive due to insufficient technical resources
helps me to discuss teaching ideas	
enhances my students learning	
provides students a wider range of experience	
provides students with a wider range of tools	

As shown in Figure 3.1, another important variable of the TAM is the participants' attitudes. Attitudes guide behavior and refer to the way an individual responds to and is disposed towards an object (Ajzen & Fishbein, 2005). This feeling or disposition may be negative or positive. The success of any initiative to implement technology in an educational program depends strongly upon the support and attitudes of the teachers involved. It has been suggested that if teachers believed or perceived computers not to be fulfilling for their own or their students' needs, they would be likely to resist any attempts to introduce any technology into their teaching–learning process (Askar & Umay, 2001). In other words, attitudes, whether positive or negative, affect how teachers respond to the technology in an instructional setting and learning environment. This in turn affects the way students view the importance of computers in schools (Teo, 2006) and affects current and future computer usage (Teo, Su Luan, & Sing, 2008). Huang and Liaw (2005) found that no matter how sophisticated and powerful the state of technology is, the extent to which it is implemented depends on teachers having a positive attitude toward it. A key reason for studying teachers' computer attitudes is the ability of attitudes to predict computer usage (Myers & Halpin, 2002). In addition, the relationship between PEU and PU on computer attitudes has been reported in various studies that provided evidence in support of a positive relationship among them (e.g., Teo, Su Luan, & Sing, 2008; Yu, Ha, Choi, & Rho, 2005). Attitudes and PU were also found to influence the individual's intention to use the technology (Davis et al., 1989).

The TAM also involves one more variable, namely, the behavioral intention (BI). The TAM implies that two behavioral beliefs, PEU and PU, have influence on an individual's intention to use technologies. In contrast to PEU and PU, which refer to process expectancy and outcome expectancy, respectively (Liaw, 2002), BI leads to actual use of technologies. The validity of this claim has been demonstrated by research evidence across a variety of contexts where technology was used (e.g., Chau, 2001; Hu, Clark, & Ma, 2003).

As mentioned above the TAM was grounded on the TRA (Ajzen & Fishbein, 1980; Fishbein, 1980). According to Sutton (1998), the TRA "has been, and continues to be, particularly popular because of its relative ability to predict behavior, simplicity, and ease of operationalization" (cited in Zint, 2002, p. 820). Albert, Aschenbrenner, and Schmalhofer (1989) characterized it as one of the most successful theories in attitude–behavior research. Similar arguments were made in previous attitudinal research studies in education (i.e., Weinburgh & Engelhard, 1994).

The TRA is based on the assumption that human beings are usually rational and make systematic use of information available to them. Therefore, people consider the implications of their actions before they decide

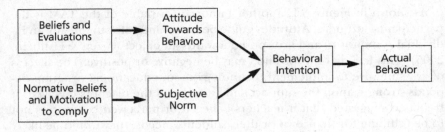

Figure 3.2 Theory of Reasoned Action (Fishbein & Ajzen, 1975).

to engage or not to engage in a given behavior (see Figure 3.2). According to Ajzen and Fishbein (1980), attitudes are a function of beliefs. Generally speaking, a person who believes that performing a given behavior will lead to mostly positive outcomes will hold a favorable attitude toward performing the behavior, while a person who believes that performing the behavior will lead to mostly negative outcomes will hold an unfavorable attitude. The beliefs that underlie a person's attitude toward the behavior are termed behavioral beliefs (Ajzen & Fishbein, 1980).

TRA further suggests that "a volitional or voluntary behavior can be predicted directly by individuals' intention to perform the behavior" (Fishbein & Ajzen, 1975, p. 288). This intention to act is a function of two determinants: "one personal in nature and the other reflecting social influence" (Ajzen & Fishbein, 1980, p. 6). The personal factor is the individual's positive or negative evaluation of performing the behavior (attitude toward the behavior). The second determinant is the person's perception of the social pressures to perform or not perform the behavior (subjective norm; see Ajzen & Fishbein, 1980). According to Ajzen and Fishbein (1973), "the relative importance of these two determinants in predicting intention to act is expected to vary with the type of behavior, situation, and based on individual differences. Variables other than attitude toward the behavior and subjective norm are assumed to influence intention to act and behavior indirectly through these two determinants" (cited in Zint, 2002, p. 824).

Despite TAM's well grounded theoretical background (TRA) and extensive and successful use across several disciplines that involve technology usage, TAM has also received severe criticism, particularly for the fact that it had only two constructs and was independent of the organizational context (Legris et al., 2003; Ma et al., 2005). Legris et al. (2003) concluded that TAM is a useful model, but has to be integrated into a broader one that would include variables related to both human and social change processes and to the adoption of the innovation model. Therefore, while for the purposes of this study we took TAM as the core framework in our study, we also looked for appropriate constructs to be put into the framework in order to provide a better understanding of the exploration of simulation technology acceptance among teachers. After reviewing relevant literature, it was found that PU could be separated in two constructs (Askar & Umay, 2001; Rotsaka, 2008): namely, the Perceived Teaching Usefulness (PTU, defined as the degree to which a teacher believes that using a particular technology will enhance his or her teaching) and the Perceived Learning Usefulness (PLU, defined as the degree to which a teacher believes that using a particular technology will enhance his or her students' learning). Additionally, it was found that the gender, the level of education received (e.g., recipient of a masters' degree), prior use of computers for educational purposes, and prior use of simulations for educational purposes were appropriate constructs to be considered in formulating the composite framework (Askar & Umay, 2001; Cox et al., 1999; Ma et al., 2005; Rotsaka, 2008; Teo, Lee, & Chai, 2008; Teo, Su Luan, & Sing, 2008; Watson, 1993; Yuen & Ma, 2002; Zacharia, 2003).

AIMS OF THE STUDY

The aims of this study were (1) to develop a questionnaire that targets teachers' beliefs, attitudes, and intentions regarding the educational use of simulations, including for science teaching purposes, and provide the findings on these constructs of the affective domain for our study's participants; (2) to explore teachers' acceptance of the educational use of simulations and to identify key intention determinants of simulation educational use; and (3) to discuss the possible contributions that our findings could have on teacher education and educational professional development programs.

MODEL AND HYPOTHESES DEVELOPMENT

As in TAM, our model, SAM (Simulation Acceptance Model), indicates that a teacher's intention to use simulation for educational purposes could be predicted and explained by his or her PEU, PTU, and PLU of simulations, in conjunction with a number of external variables, namely, gender, prior training on computer use, and prior experience with computer and simulation use for educational purposes (see Figure 3.3).

As shown in Figure 3.3, our model SAM used TAM for its theoretical basis. Specifically, the PEU, attitudes, behavioral intentions, and actual use are the constructs that remained the same. The PU was retained, but it was replaced by two new constructs, the PLU and PTU. Given the proposed SAM of Figure 3.3, we posited the following hypotheses:

H1: *The gender of the teacher influences his or her PEU of simulations.*

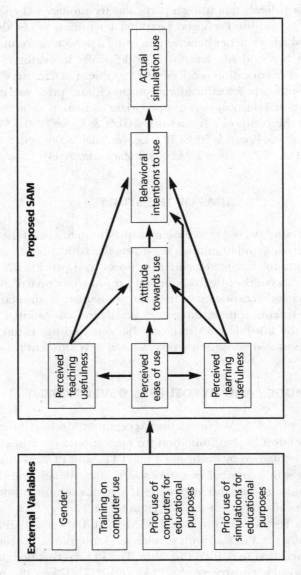

Figure 3.3 The proposed Simulation Acceptance Model.

H2a: *A teacher's prior use of computers for educational purposes has a positive effect on his or her PEU of simulations.*

H2b: *A teacher's prior use of computers for educational purposes has a positive effect on his or her PTU of simulations.*

H2c: *A teacher's prior use of computers for educational purposes has a positive effect on his or her PLU of simulations.*

H3a: *A teacher's prior use of simulations for educational purposes has a positive effect on his or her PEU of simulations.*

H3b: *A teacher's prior use of simulations for educational purposes has a positive effect on his or her PTU of simulations.*

H3c: *A teacher's prior use of simulations for educational purposes has a positive effect on his or her PLU of simulations.*

H4a: *A teacher's prior training on computer use has a positive effect on his or her PEU of simulations.*

H4b: *A teacher's prior training on computer use has a positive effect on his or her PTU of simulations.*

H4c: *A teacher's prior training on computer use has a positive effect on his or her PLU of simulations.*

H5a: *A teacher's PEU of simulations has a positive effect on his or her PTU of simulations.*

H5b: *A teacher's PEU of simulations has a positive effect on his or her PLU of simulations.*

H5c: *A teacher's PEU of simulations has a positive effect on his or her attitudes towards the use of simulations.*

H5d: *A teacher's PEU of simulations has a positive effect on his or her BI.*

H6a: *A teacher's PTU of simulations has a positive effect on his or her attitudes towards the use of simulations for educational purposes.*

H6b: *A teacher's PTU of simulations has a positive effect on his or her BI to accept the use of simulations for educational purposes.*

H7a: *A teacher's PLU of simulations has a positive effect on his or her attitudes towards the use of simulations for educational purposes.*

H7b: *A teacher's PLU of simulations has a positive effect on his or her BI to accept the use of simulations for educational purposes.*

H8: *A teacher's attitude towards the use of simulations for educational purposes has a positive effect on his or her BI to accept the use of simulations for educational purposes.*

H9: *Attitudes can be integrated in a structural equation model with PEU, PTU, PLU, and BI.*

H10: *PTU and PLU can be integrated at the same time in a structural equation model with PEU and BI.*

H11: *PEU is a determinant of BI in a structural equation model.*

H12: *PEU is a determinant of PTU in a structural equation model.*

H13: *PEU is a determinant of PLU in a structural equation model.*

Finally, by using K-means clustering, we explored the possibility of distinguishing among sub-samples of respondents due to different responses to questionnaire items. In this regard, we investigated the heterogeneity of the sample in terms of their dispositions towards simulations.

METHODS

Sample

The sample of the study comprised of 514 in-service elementary school teachers from Cyprus (342 Females and 172 Males; Mean$_{age}$ = 31.2, SD = 6.8), who were aware of what a simulation is. For this purpose we provided the subjects through the study's questionnaire with a definition and a picture of a simulation. The reasoning behind the provision of a definition was to ensure that all participants were referring to the same construct. All teachers were randomly selected from across the country. However, they were included in the study's sample only if they were owners of a computer and had taught science classes. Usually, elementary school teachers in Cyprus teach more than one of the subject domains of the national curriculum (e.g., science, mathematics, language, technology, geography). Finally, data were collected using a user-reported self-assessment approach, deemed appropriate because of considerable literature support for its use in intention-based studies (e.g., Davis, 1989). The background of the respondents is summarized in Table 3.3. One third of respondents were male (33.5%). A considerable percentage (43.6%) had teaching experience of five years or less. Most participants had completed post-graduate studies (62.1%), and an analogous majority served in urban school districts (60.8%). About one fifth of the sample taught 1st or 2nd grade (18.3%), while 38.1% taught 3rd or 4th grade, and another 43.7% taught 5th or 6th grade. All teachers received prior training in computer use, and almost all of them had used computers in their classes (95.9%). A substantial number had also used simulations in their classes (72.4%). Simulation use pertained to science classes for 54.3% of the respondents.

TABLE 3.3 Background of Respondents

Variables	Categories	Percentage
Gender	Male	33.5
	Female	66.5
Teaching experience	0–5 years	43.6
	6–10 years	25.9
	11–15 years	14.0
	> 15 years	16.5
Education level	Bachelor's degree	37.9
	Post-graduate studies	62.1
School district	Rural	39.2
	Urban	60.8
Grade	1st and/or 2nd	18.3
	3rd and/or 4th	38.1
	5th and/or 6th	43.7
Prior training on computer usage	Yes	100.0
	No	0.0
Computer use for educational purposes	Yes	95.9
	No	4.1
Simulation use for educational purposes	Yes	72.4
	No	27.6
Science classes	Yes	54.3
	No	45.7

Instrumentation

Special attention to explaining "what is a simulation" was a critical piece for a valid and successful completion of the instrument. Thus, we developed a survey instrument that included a definition of a simulation and a picture with an example of a simulation. The definition was piloted, reviewed, and refined and validated (deemed to be appropriate) by a five-member expert panel of researchers.

The questionnaire, after it was piloted, reviewed, refined and validated, consisted of 59 items that were separated in six parts (see Appendix). The first part included nine items that focused on collecting demographic data (e.g., gender, age, years of teaching experience), information on prior training on computer use, and information on prior experience with use of computers and simulations for educational purposes. The second part involved five closed-ended items that measured participants' beliefs about the PEU of simulations. The third part involved nine closed-ended items that measured participants' beliefs about the PTU of simulations. The fourth part involved 18 closed-ended items that measured participants' beliefs about the PLU of simulations. The fifth part involved nine closed-ended items that measured participants' attitudes towards the educational use of simulations. Lastly, the sixth part involved nine closed-ended items that measured participants' behavioral intentions to the educational use of simulations. The items of the second, third, fourth, fifth, and sixth parts were accompanied by a numerical Likert scale. Participants gave their opinions to each statement on a five-point Likert scale, ranging from 1 (strongly disagree) to 5 (strongly agree).

All items were derived from prior research work: second part (Christensen & Knezek, 1996; Davis, 1989; Rotsaka, 2008), third and fourth parts (Askar & Umay, 2001; Davis, 1989; Rotsaka, 2008; Zacharia, 2003), fifth part (Christensen & Knezek, 1996; Fishbein & Azjen, 1975; Zacharia, 2003), and sixth part (Christensen & Knezek, 1996; Davis, 1989; Rotsaka, 2008). All items were slightly modified to refer to simulations. The original pool of items included 76 items. All items were piloted, reviewed, and refined by a seven-member expert panel (3 science educators, 3 psychologists, and 1 sociologist) that determined the final items that measured consistently the various constructs. Item reliability scores and factor analytic data revealed the 59 final five Likert scale items that measure consistently and parsimoniously the various constructs. In short, all of the constructs involved in the questionnaire (e.g., PEU, PTU, PLU, AT, BI) produced high reliability scores and clean factor solutions (see Table 3.4).

TABLE 3.4   Factor Structure of Questionnaire Items

	Components				
	1	**2**	**3**	**4**	**5**
Perceived ease of use (PEU)					
PEU1	0.50	0.38	0.38	0.20	**0.52**
PEU2	0.48	0.44	0.35	0.19	**0.52**
PEU3	0.41	0.34	0.39	0.20	**0.58**
PEU4	−0.11	0.31	0.46	0.06	**0.57**
PEU5	0.17	0.12	0.17	0.06	**0.81**
Perceived teaching usefulness (PTU)					
PTU1	**0.57**	0.43	0.10	0.35	0.18
PTU2	**0.67**	0.36	0.11	0.22	0.21
PTU3	**0.52**	0.38	0.27	0.19	0.47
PTU4	**0.70**	0.22	0.17	0.27	0.28
PTU5	**0.59**	0.35	0.25	0.27	0.32
PTU6	**0.74**	0.36	0.13	0.25	0.21
PTU7	**0.71**	0.31	0.11	0.23	0.27
PTU8	**0.54**	0.47	0.37	0.19	0.28
PTU9	**0.65**	0.31	0.17	0.38	0.25

TABLE 3.4 Factor Structure of Questionnaire Items (cont.)

	Components				
	1	2	3	4	5
Perceived learning usefulness (PLU)					
PLU1	0.45	**0.57**	0.10	0.41	0.12
PLU2	0.53	**0.65**	0.17	−0.02	0.26
PLU3	0.47	**0.54**	0.32	−0.10	0.41
PLU4	0.30	**0.77**	0.29	0.19	−0.01
PLU5	0.47	**0.56**	0.25	0.21	−0.08
PLU6	0.23	**0.72**	0.34	0.14	0.27
PLU7	0.36	**0.63**	0.34	0.25	0.33
PLU8	0.48	**0.61**	0.18	0.25	0.26
PLU9	0.31	**0.73**	0.13	0.32	0.03
PLU10	0.40	**0.67**	0.32	0.09	0.03
PLU11	0.36	**0.71**	0.19	0.31	−0.04
PLU12	0.32	**0.56**	0.28	0.13	0.52
PLU13	0.28	**0.74**	0.35	0.09	0.17
PLU14	0.47	**0.58**	0.26	−0.07	0.16
PLU15	0.38	**0.69**	0.39	0.00	0.10
PLU16	0.50	**0.69**	0.27	0.13	0.02
PLU17	0.38	**0.69**	0.26	0.05	0.30
PLU18	0.38	**0.67**	0.30	0.01	0.24
Attitudes towards the educational use of simulations (AT)					
AT1	0.29	0.13	**0.79**	−0.03	−0.07
AT2	0.32	0.34	**0.65**	−0.05	0.07
AT3	0.13	0.15	**0.74**	−0.06	0.01
AT4	0.11	0.33	**0.85**	0.01	−0.01
AT5	0.04	0.30	**0.84**	0.04	0.13
AT6	0.13	0.26	**0.83**	0.02	0.02
AT7	0.25	0.49	**0.66**	−0.03	0.07
AT8	0.17	0.43	**0.71**	0.00	−0.10
AT9	0.31	0.54	**0.61**	0.01	0.06
Behavioral intentions to the educational use of simulations (BI)					
BI1	0.24	0.31	0.05	**0.83**	−0.07
BI2	0.25	0.34	0.04	**0.81**	−0.06
BI3	0.16	0.22	0.21	**0.83**	−0.02
BI4	0.05	0.35	0.03	**0.73**	0.03
BI5	0.22	0.15	0.01	**0.83**	0.10
BI6	0.14	0.39	0.10	**0.75**	0.13
BI7	0.17	0.11	0.15	**0.80**	0.23
BI8	0.14	0.15	0.06	**0.85**	0.12
BI9	0.00	0.12	0.19	**0.66**	0.41
Variance explained	57.36	66.34	70.42	73.66	75.85
Cronbach's alpha	0.95	0.97	0.95	0.94	0.92

Note: Maximum factor loadings for each item across factors are presented in bold.

Data Analysis

The data analysis served three different goals. The first one concerned analyses that targeted the validity and reliability checks of our instrument. The second one tested the proposed research model (SAM), and this involved assessing the contributions and significance of the manifest variables path co-efficients. The third one concerned the analysis of the data collected from our study's participants. The idea was to examine what beliefs, attitudes, and intentions our participants held towards the educational use of simulations.

All of the closed-ended items (Likert-based items) of the questionnaire's parts 2 through 6 were entered into SPSS sheets according to the selection that each teacher made on the numerical Likert scale that accompanied each item. The analysis involved the calculation of means for each part of the questionnaire and for each participant separately. Participants were designated as having a positive, neutral, or negative belief, attitude, or intention based on the average value of the answers they gave for the items in each part of the instrument: The negative range was between 1 and 2.5 (2.5 is included), the neutral range was between 2.5 and 3.5 (3.5 is not included) and the positive range was between 3.5 and 5 (5 is included). The validity of our instrument was tested by means of factor analysis. Cronbach's alpha was computed for each factor to estimate the internal consistency of each construct. Analysis of variance was used to examine the effect of external variables on various constructs (i.e., PEU, PTU, PLU). Pearson correlation coefficients were computed to investigate interrelations among components (i.e., PEU, PTU, PLU, AT, BI). We conducted a k-means cluster analysis to examine differences among respondents in terms of their responses to questionnaire items. Pearson correlations were computed for the entire sample and sub-samples derived by k-means clustering. We also conducted cross-tabulations to investigate significant trends of cluster membership of respondents with their background variables. The testing of the model was executed by means of structural equation modeling.

RESULTS

Validity and Reliability of the Instrument

We employed principal component analysis extraction method using varimax with Kaiser normalization rotation (Kaiser-Meyer-Olkin Measure of Sampling Adequacy = 0.952). The factor analysis retained five components with eigenvalues greater than 1. The factor loadings of all individual items exceeded the 0.50 threshold set by Hair, Black, Babin, Anderson, and Tatham (2006) and ranged from 0.52–0.85 (Table 3.4). Since factor loadings among items within the same component were high and, at the same time, factor loadings across components were low; all constructs exhibited both convergent and discriminate validity, respectively. In this regard, factor analysis confirmed the corresponding constructs that were hypothesized in the model. The cumulative percentage of variance explained by the five factors amounted to 75.85%. Internal consistency was examined using Cronbach's alpha values. As shown in Table 3.4, the reliabilities of components were between 0.92 and 0.97, which surpassed the 0.70 acceptance limit suggested by DeVellis (2003).

Descriptive Statistics of Questionnaire Items

Means, standard deviations, and median values for questionnaire items are presented for each factor in Tables 3.5–3.9. All median values were equal to 5, while means ranged from 4.31 to 4.73. These findings indicate an extended endorsement of simulations by elementary school teachers. No standard deviation was greater than one, which shows that respondent replies were dispersed close to average values.

TABLE 3.5   Descriptive Statistics for PEU

Items	Median	Mean	Standard deviation
Interacting with simulations is simple	5	4.44	0.75
A simulation is a tool that can easily be used	5	4.31	1.00
The use of simulations is simple	5	4.43	0.76
Learning to operate simulations is easy for me	5	4.34	0.91
It is easy for me to become skillful in using simulations	5	4.33	0.96

TABLE 3.6   Descriptive Statistics for PTU

Items	Median	Mean	Standard deviation
Simulations for a fact can help me during my teaching	5	4.53	0.76
Simulations make my teaching easier	5	4.52	0.74
Simulations provide direct experiences with phenomena	5	4.60	0.65
Working with simulations makes teaching more interesting	5	4.73	0.52
Simulations relieve teachers of routine duties	5	4.57	0.69
Simulations may improve the overall quality teaching	5	4.64	0.64
Simulations could provide unique visualizations	5	4.71	0.57
Simulations make teaching more enjoyable	5	4.41	0.81
Simulations are a useful tool for teachers	5	4.59	0.65

TABLE 3.7 Descriptive Statistics for PLU

Items	Median	Mean	Standard deviation
The use of simulations helps students acquire skills	5	4.62	0.64
Simulations enhance the understanding of complex concepts	5	4.61	0.65
The use of simulations enhances conceptual understanding	5	4.50	0.76
Simulations can support student group working	5	4.36	0.91
The use of simulations promotes the use of learning strategies (e.g., problem solving)	5	4.60	0.64
Simulations stimulate students' creativity	5	4.34	0.94
The use of simulations provides a better learning experience to the students than a traditional mode of instruction	5	4.57	0.65
The use of simulations helps students to give better explanations	5	4.54	0.67
Simulations actively engage students during a learning activity	5	4.55	0.73
The use of simulations increases interest towards learning	5	4.62	0.58
The use of simulations promotes the active participation of all students	5	4.59	0.72
The use of simulations gives the opportunity to manipulate all variables associated with the phenomenon under study	5	4.56	0.77
The use of simulations accommodates all students' different needs	5	4.33	0.96
Simulations make the presentation of all concepts (both real and reified) "observable" to the students	5	4.69	0.55
Simulations help students to self-regulate their own learning	5	4.37	0.90
Simulations improve students' achievement	5	4.44	0.76
The use of simulations promotes students' problem solving skill	5	4.47	0.75
The use of simulations develops students' critical thinking	5	4.46	0.79

TABLE 3.8 Descriptive Statistics for "Attitude Towards the Educational Use of Simulations"

Items	Median	Mean	Standard deviation
Simulations confuse me	5	4.51	0.90
I like using simulations in my teaching	5	4.31	1.00
I feel intimidated when I have to use simulations	5	4.54	0.85
The use of simulation is boring	5	4.68	0.58
The use of simulation disappoints me	5	4.69	0.63
The use of simulation wares me out	5	4.62	0.70
The use of simulation excites me	5	4.34	0.91
The use of simulation interest me little	5	4.62	0.68
Working with simulations makes my teaching enjoyable	5	4.55	0.68

Note: Items 1, 3, 4, 5, 6 and 8 were reverse coded.

TABLE 3.9 Descriptive Statistics for "Behavioral Intentions to the Educational Use of Simulations"

Items	Median	Mean	Standard deviation
I will try to learn anything that relates with simulations and their educational use	5	4.61	0.70
I would like to participate in seminars/courses that focus on the educational use of simulations	5	4.63	0.68
I would like to take simulation courses	5	4.57	0.81
I will pursuit to participate in courses of other teachers which use simulations during their teaching	5	4.43	0.93
I will like to be informed on recent developments in the educational use of simulations	5	4.62	0.67
I will teach courses that favor the use of simulations	5	4.34	0.94
I will use simulations as little as possible in my teaching	5	4.66	0.64
I will use simulations in my teaching, if I have the right equipment	5	4.62	0.66
I intend to use simulations in my teaching	5	4.58	0.87

Note: Item 7 was reverse coded

The Effect of the External Variables on SAM's Constructs

Males (mean = 4.29) presented significantly higher mean value of PEU compared to females (mean = 4.04), (F = 20.11, $p < 0.001$), which supports the hypothesis H1. Moreover, computer use as well as simulation use were found to have a significant positive effect on teachers' PEU (F = 20.38, $p < 0.001$, and F = 215.80, $p < 0.001$, respectively), PTU (F = 19.37, $p < 0.001$, and F = 109.49, $p < 0.001$, respectively), and PLU (F = 16.50, $p < 0.001$, and F = 85.78, $p < 0.001$, respectively). These findings confirmed hypotheses H2a, H2b, H2c, H3a, H3b, and H3c. On the other hand, the prior training on computer use variable was not found to influence teachers' PEU, PTU and PLU, which resulted in rejecting the hypotheses H4a, H4b, and H4c.

Correlations Among Constructs

Correlations among factors are presented in Table 3.10. Pearson correlation coefficients were highly significant for all cases. These results support hypotheses H5a, H5b, H5c, H5d, H6a, H6b, H7a, H7b, and H8.

Heterogeneity Among Respondents

Our next aim was to investigate whether respondents could be distinguished in sub-samples on the basis of their responses to questionnaire

TABLE 3.10   Pearson Correlations Among Factors

	Perceived teaching usefulness	Perceived learning usefulness	Attitudes towards the educational use of simulations	Behavioral intentions to the educational use of simulations
Perceived ease of use	0.77	0.78	0.65	0.67
Perceived teaching usefulness		0.86	0.83	0.53
Perceived learning usefulness			0.79	0.64
Attitudes towards the educational use of simulations				0.54

Note: All Pearson correlations are significant at the $p < 0.001$ significance level.

TABLE 3.11   K-means Clustering of Respondents by Factor Scores

Factors	Cluster 1 (58.75% of the sample)	Cluster 2 (41.25% of the sample)	F
Perceived ease of use	4.55	3.52	1313.94
Perceived teaching usefulness	4.97	4.04	933.62
Percived learning usefulness	4.96	3.87	1417.39
Attitudes towards the educational use of simulations	4.97	3.90	875.10
Behavioral intentions to the educational use of simulations	4.95	4.01	510.28

Note: Factor scores have been computed as average recording across items for each factor; all F values are significant at $p < 0.001$.

items. We calculated averages across items for each factor and subjected these values to k-means clustering. We selected a two-cluster solution, which distinguished among respondents through the highest number of factors. The results of the analysis are shown in Table 3.11. A sub-sample of elementary school teachers (58.75%) presented significantly higher average values for all factors compared to another sub-sample (41.25%) that did not seem to endorse simulations to the same extent. Differences were most pronounced in the case of "perceived learning usefulness" and least marked in "intention to use."

In order to investigate whether respondent clusters differed in terms of correlations between factors, we computed Pearson correlations among average values for factors in respondents' cluster 1 (Table 3.12) and cluster 2 (Table 3.13). Respondents' cluster 1 (Table 3.12) followed the pattern of the entire sample (Table 3.10). Once again, all Pearson correlation coefficients

TABLE 3.12 Pearson Correlations Among Factors for Respondents' Cluster 1

	Perceived teaching usefulness	Perceived learning usefulness	Attitudes towards the educational use of simulations	Behavioral intentions to the educational use of simulations
Perceived ease of use	0.60	0.31	0.39	0.87
Perceived teaching usefulness		0.50	0.51	0.54
Perceived learning usefulness			0.40	0.44
Attitudes towards the educational use of simulations				0.38

Note: All Pearson correlations are significant at the $p < 0.001$ significance level.

TABLE 3.13 Pearson Correlations Among Factors for Respondents' Cluster 2

	Perceived teaching usefulness	Perceived learning usefulness	Attitudes towards the educational use of simulations	Behavioral intentions to the educational use of simulations
Perceived ease of use	0.25***	0.15*	−0.15*	0.08ns
Perceived teaching usefulness		0.59***	0.53***	−0.13ns
Perceived learning usefulness			0.34***	0.04ns
Attitudes towards the educational use of simulations				−0.10ns

Note: ns = non significant; * $p < 0.05$; ** $p < -.01$; *** $p < 0.0001$.

were significant at the $p < 0.001$ level. Correlations for cluster 2 (Table 3.13) revealed important differences from the pattern of correlations that was found for the entire sample and respondents' cluster 1. For members of cluster 2, who presented significantly lower average values across all factors, there was no significant correlation of any factor with "intention to use."

Further, respondent clusters also differed in terms of a series of background variables (Table 3.14). In particular, males (Likelihood ratio $\chi^2 = 17.81$, $p < 0.001$), respondents with more teaching experience (Likelihood ratio $\chi^2 = 43.70$, $p < 0.001$), teachers with post-graduate studies (Likelihood ratio $\chi^2 = 84.85$, $p < 0.001$), subjects in urban school districts (Likelihood ratio $\chi^2 = 48.78$, $p < 0.001$), those who teach at higher grades (Likelihood ratio $\chi^2 = 10.99$, $p < 0.01$), and participants who use computers

TABLE 3.14 Cross Tabulations of K-Means Cluster Membership by Gender, Teaching Experience, Education Level, School District, Grade, Computer Use, Simulation Use, and Use of Simulations in Science Classes

Variables	Categories	Cluster 1	Cluster 2	Likelihood ratio χ^2	Cramer's V
Gender	Male	71.5	28.5	17.81***	0.18***
	Female	52.3	47.7		
Teaching experience	0–5 years	43.3	56.7	43.70***	0.29***
	6–10 years	64.7	35.3		
	11–15 years	76.4	23.6		
	> 15 years	75.3	24.7		
Education level	Bachelor's degree	33.3	66.7	84.85***	0.40***
	Post-graduate studies	74.3	25.7		
School district	Rural	39.5	60.5	48.78***	0.31***
	Urban	70.6	29.4		
Grade	1st and/or 2nd	43.5	56.5	10.99**	0.16**
	3rd and/or 4th	60.5	39.5		
	5th and/or 6th	64.5	35.5		
Computer use	Yes	61.3	38.7	38.48***	0.25***
	No	0.0	100.0		
Simulation use	Yes	73.1	26.9	117.21***	0.47***
	No	21.1	78.9		
Science classes	Yes	59.9	40.1	0.31ns	0.02ns
	No	57.4	42.6		

Note: Numbers presented correspond to percentages within variable categories; ns = non significant; * $p < 0.05$; ** $p < 0.01$; *** $p < 0.001$.

(Likelihood ratio χ^2 = 38.48, $p < 0.001$) or simulations in their classes (Likelihood ratio χ^2 = 117.21, $p < 0.001$) were more likely to belong to cluster 1 and endorse simulations more.

Test of the Proposed Model

STATISTICA was used to perform structural equation modeling and test the fit between the hypothesized model and the obtained data. We used covariance matrices and employed maximum likelihood estimation. Model fit was assessed by the ratio of the χ^2 statistic to its degree of freedom and two other indices, namely the Jöreskog goodness of fit index (GFI) and the Root Mean Square (RMS) standardized residual (Kline, 2005). The ratio of the χ^2 statistic to its degree of freedom should be less than 5 (Hair et al., 2006). Acceptable model fit presupposes a value of Jöreskog GFI greater than 0.90 and a value of RMS standardized residual lower than 0.05.

The first tests showed that no solution could be acceptable in terms of fit indices if "attitudes towards the educational use of simulations" were included in the model. Therefore, we had to reject hypothesis H9. Moreover, PTU and PLU could not coexist in the model for the same reason; that is, when they coexisted, the solution did not present acceptable model fit. Based on this finding, we rejected hypothesis H10.

As a next step, we tested two alternative versions of a subset of the initial model. In both cases, the ratio of the χ^2 statistic to its degree of freedom was well below the recommended value of 5. Standardized parameter estimates for the two alternative solutions, which converged and presented acceptable fit induces, are presented in Figures 3.4 and 3.5. The first model (see Figure 3.4) involved PTU, PEU, and BI (Jöreskog GFI = 0.95; RMS standardized residual = 0.026), while the second model (see Figure 3.5) included PLU, PEU, and BI (Jöreskog GFI = 0.95; RMS standardized residual = 0.034). BI was predicted in both solutions by PEU ($\beta = .64$, $p < 0.001$ for model 1; $\beta = .63$, $p < 0.001$ for model 2). These results confirmed hypothesis H11. Furthermore, it was found that PEU influenced PTU in the

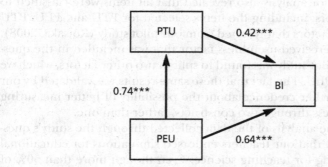

Figure 3.4 Structural equation modelling testing results for model 1.

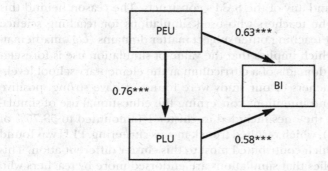

Figure 3.5 Structural equation modelling testing results for model 2.

first resulting model ($\beta = .74$, $p < 0.001$) and PLU in the second resulting model ($\beta = .76$, $p < 0.001$). These findings support H12 and H13, respectively. PTU and PLU predicted BI in the first ($\beta = .42$, $p < 0.001$) and second model ($\beta = .58$, $p < 0.001$), respectively.

DISCUSSION

The objectives of this study were (1) to develop a questionnaire that targets teachers' beliefs, attitudes, and intentions regarding the educational use of simulations, including for science teaching purposes, and provide the findings on these constructs of the affective domain for our study's participants; (2) to explore teachers' acceptance of the educational use of simulations and to identify key intention determinants of simulation educational use; and (3) to discuss the possible contributions that our findings could have on teacher education and educational professional development programs.

As far as the first objective is concerned, we managed to develop a questionnaire whose constructs produced high reliability scores and clean factor solutions. The factor analysis also revealed that all items were classified to pre-specified factors, including the items selected for PTU and PLU. PTU and PLU are two factors that resulted from our pilot study (Rotsaka, 2008). Specifically, the perceived usefulness factor that was included in the questionnaire of our pilot study was found to split in two other factors, which we named PTU and PLU. The fact that these same results was validated by our study, provides further credence about the possibility of better measuring perceived usefulness through two constructs, rather than one.

Additionally, the analysis of the data collected through the study's questionnaire revealed that our teachers endorsed simulations for educational purposes, including for teaching science. Even though more than 50% of our teachers used simulations for teaching science (most prevalent subject domain for simulation use), no significant relationship was found between science teaching and any of the SAM's constructs. The reason behind this finding was that the teachers who used simulations for teaching science also used them for teaching other subject matter domains (e.g., mathematics, geography), which implies that the value of simulation use is foreseen across the subject domains of a curriculum at the elementary school level.

Overall, the teachers of our study were found to have strong, positive beliefs, attitudes and intentions concerning the educational use of simulations, particularly, the ones included in cluster 1 (amounted to 58.75% of the study's sample), which resulted from k-means clustering. PLU was found to be the factor which contributed most to this cluster differentiation. This latter finding implies that simulations are endorsed more by teachers who strongly believe that simulations contribute to their students' learning. The fact that our analysis showed our teachers to highly endorse the use of simulations for educational purposes is quite encouraging, given that recent research has shown the positive influence that the use of simulations has on student learning (Triona & Klahr, 2003; van der Meij & de Jong, 2006; Zacharia & Anderson, 2003; Zacharia, 2007; Zacharia et al., 2008).

Moreover, the conduction of cross tabulations of k-means cluster membership by gender, teaching experience, education level, school district, grade, computer use, simulation use, and use of simulations in science classes, revealed that the group of respondents who endorsed simulations more included (a) more males, (b) teachers with more years of teaching experience, (c) teachers with post-graduate studies, (d) teachers working in urban districts, (e) teachers teaching at higher grades, and (f) teachers who used computers or simulations in their classes. These findings are important because they provide us with a picture in terms of which teachers we should target in order to increase future use of simulations in schools. Needless to say, these findings also have obvious implications on teacher education and educational professional development programs.

In terms of the second objective, our findings on SAM were found to be consistent with previous research findings on TAM (Davis, 1989; Davis et al., 1989; Venkatesh, 1999; Venkatesh & Davis, 1996, 2000). Specifically, they confirmed the positive influence of PEU and PU (PTU and PLU) on BI, as well as the positive influence of PEU on PU (PTU and PLU). On the other hand, it was found that attitudes could not be integrated in a model with PEU, PTU, PLU, and BI. This latter finding is not in line with previous studies using TAM in predicting intention and usage (Teo, Lee, & Chai, 2008; Venkatesh et al., 2003). Another important finding that relates to the resulting SAM is that PTU and PLU could not be integrated together in the same model. As a result, we ended up with two models, rather than just one (see Figures 3.4 and 3.5), with the one that includes the PLU having the strongest significant direct positive effect on BI.

As shown in Figures 3.4 and 3.5, besides PLU and PTU, PEU was also found to be a determinant in the model in predicting intention to use. Furthermore, PEU was found to be a determinant in the model in predicting both PTU and PLU. These findings seem to indicate that the PLU, PTU, and PEU among teachers are extremely important.

As far as the third objective of the study is concerned, it appears that it is important to emphasize in teacher education and educational professional development programs the benefits that the use of simulations offers to our teaching and to our students' learning. For example, we could expose teachers to good practices that involve the use of simulations and, hence, show them how learning goals could be more efficiently and effectively achieved through the use of simulations. Given that a considerable number of teachers do not consider computer technology as a pedagogical instrument (Ma et al., 2005), exposing teachers to the teaching and learning benefits of the use of simulations becomes imperative. Additionally, teachers need the necessary knowledge and competence in order to use simulations. This means that teacher education and educational professional development programs should expose teachers to simulation use in order to understand that simulations are easy to use.

While acknowledging the implications of SAM to teacher education and teacher development, we would also like to address the issue of applicability of SAM in examining simulation use for educational purposes. Although the results from the data analysis were generally good, there are obvious limitations to the present study. For instance, it is highly probable that our attempt to produce extensive scales in terms of capturing a wide array of possible dimensions of the constructs under study led to increased complexity, which could not be handled in the case of demanding statistical approaches, such as structural equation modeling. However, given the good model-fit of the final models, we argue that the final models give a good understanding of teacher acceptance of the use of simulations for educational purposes.

REFERENCES

Agarwal, R., & Prasad, J. (1999). Are individual differences germane to the acceptance of new information technologies? *Decision Sciences, 30*(2), 361–391.

Ajzen, I. & Fishbein, M. (1973). Attitudinal and normative variables as predictors of specific behaviors. *Journal of Personality and Social Psychology, 27,* 41–57.

Ajzen, I. & Fishbein, M. (1980). *Understanding attitudes and predicting social behavior.* Englewood Cliffs, NJ: Prentice Hall.

Ajzen, I., & Fishbein, M. (2005) The influence of attitudes on behavior. In D. Albarracín, B. T. Johnson & M. P. Zanna (Eds.), *Handbook of attitudes and attitude change: Basic principles* (pp. 173–221). Mahwah, NJ: Erlbaum.

Albert, D., Aschenbrenner, K.M., & Schmalhofer, F. (1989). Cognitive choice processes and the attitude–behavior relation. In A. Upmeyer (Ed.), *Attitudes and behavioral decisions* (pp. 61–99). New York: Springer-Verlag.

Askar, P. & Umay, A. (2001). Pre-service elementary mathematics teachers' computer self-efficacy, attitudes towards computers, and their perceptions of computer-enriched learning environments. In *Proceedings of Society for Information Technology and Teacher Education International Conference* (pp. 2262–2263). Chesapeake, VA: AACE.

Barki, H. & Hartwick, J. (1994). Measuring user participation, user involvement, and user attitude. *MIS Quarterly, 1,* 59–78.

Becker, H. (2001). How are teachers using computers in instruction? *Paper presented at the 2001 Meetings of the American Educational Research Association,* Retrieved 25 August 2008 from: http://www.crito.uci.edu/tlc/findings/conferences-pdf/how_are_teachers_using.pdf.

Bencze, L. & Hodson, D. (1999). Changing practice by changing practice: Toward more authentic science and science curriculum development. *Journal of Research in Science Teaching, 36*, 521–539.

Bereiter, C., & Scardamalia, M. (2006). Education for the knowledge age. In P. A. Alexander, & P. H. Winne (Eds.), *Handbook of educational psychology* (2nd Ed., pp. 695–713). Mahwah, NJ : Lawrence Erlbaum.

Breen, R, Lindsay, R., Jenkins A., & Smith P. (2001). The role of information and communication technologies in a university learning environment. *Studies in Higher Education, 26*, 95–114.

Burns, M. (2002). From black to white to color: technology, professional development and changing practice. *T.H.E. Journal online*, Retrieved from http://thejournal.com/articles/16020.

Chau, P. Y. K. (2001). Influence of computer attitude and self-efficacy on IT usage behaviour. *Journal of End-User Computing, 13*(1), 26–33.

Christensen, R. & Knezek, G. (1996). *Constructing the teachers' attitudes toward computers (TAC) questionnaire.* Paper presented to the Southwest Educational Research Association Annual Conference, New Orleans, LA, January. Retrieved from http://www.tcet.unt.edu/research/survey/tac222.pdf

Cox, M.J., Preston, C., & Cox, K. (1999). *What motivates teachers to use ICT?* Paper presented at the British Educational Research Association Conference, Brighton, UK, September.

Culpan, O. (1995). Attitudes of end-users towards information technology in manufacturing and service industries. *Information and Management, 28*, 167–176.

Davis, F.D. (1989). Perceived usefulness, perceived ease of use and user acceptance of information technology. *MIS Quarterly, 13*(3), 319–339.

Davis, F. D., Bagozzi, R. P., & Warshaw, P. R. (1989). User acceptance of computer technology: a comparison of two theoretical models. *Management Science, 35*, 982–1003.

DeVellis, R. F. (2003). *Scale development: Theory and applications* (2nd Ed.). Newbury Park, CA: Sage.

Dishaw, M. T., & Strong, D. M. (1999). Extending the technology acceptance model with task-technology fit constructs. *Information & Management, 36*(1), 9-21.

Drennan, J., Kennedy, J., & Pisarksi, A. (2005). Factors affecting student attitudes toward flexible online learning in management education. *The Journal of Educational Research*, 98, 331–340.

Fishbein, M. (1980). A theory of reasoned action: Some applications and implications. In Howe, H. & Page, M. (Eds.), *Nebraska symposium on motivation* (pp. 65–116). Lincoln:
University of Nebraska Press.

Fishbein, M. & Ajzen, I. (1975). *Belief, attitude, intention, and behavior: An introduction to theory and research.* Reading, MA: Addison-Wesley.

Gao, Y. (2005). Applying the technology acceptance model (TAM) to educational hypermedia: A field study. *Journal of Educational Multimedia and Hypermedia, 14*(3), 237–247.

Hair, J. F. Jr., Black, W. C., Babin, B. J., Anderson, R. E., & Tatham, R. L. (2006). *Multivariate data analysis* (6th Ed.). Englewood Cliffs, NJ: Prentice-Hall.

Hu, P. J., Clark, T. H. K., & Ma, W. W. (2003) Examining technology acceptance by school teachers: A longitudinal study. *Information & Management, 41*(2), 227–241.

Huang, H. M., & Liaw, S. S. (2005). Exploring user's attitudes and intentions toward the web as a survey tool. *Computers in Human Behavior, 21*(5), 729–743.

Jackson, C.M., Chow, S., & Leitch, R.A. (1997). Toward an understanding of the behavioral intention to use an information system. *Decision Sciences, 28*(2), 357–389.

Kay, R. H. (1992). An analysis of methods used to examine gender differences in computer-related behavior. *Journal of Educational Computing Research, 8*(3), 277–290.

Kay, R. H. (1993). An exploration of theoretical and practical foundations for assessing attitudes toward computers: the Computer Attitude Measure (CAM). *Computers in Human Behavior, 9,* 371–386.

Kline, R. B. (2005). *Principles and practice of structural equation modelling* (2nd Ed.). New York: Guilford Press.

Kluever, R.C., Lam, T.C.M., & Hoffman, E.R. (1994). The computer attitude scale: Assessing changes in teachers' attitudes toward computers. *Journal of Educational Computing Research, 11,* 251–256.

Koohang, A.A. (1989). A study of attitudes toward computers: Anxiety, confidence, liking and perception of usefulness. *Journal of Research on Computing in Education, 22,* 137–150.

Koslowsky, M., Holman, M., & Lazar, A. (1990). Predicting behavior on a computer from intentions, attitudes, experience. *Current Psychology: Research and Reviews, 9,* 75–83.

Lawton, J., & Gerschner, V.T. (1982). A review of the literature on attitudes towards computers and computerized instruction. *Journal of Research and Development in Education, 16*(1), 50–55.

Legris, P., Ingham, J., & Collerette, P. (2003). Who do people use information technology? A critical review of the technology acceptance model. *Information & Management, 40,* 191–204.

Liaw, S.S. (2002). Understanding user perceptions of world-wide web environments. *Journal of Computer Assisted Learning, 18*(2), 137–148.

Ma, W.W.K., Andersson, R., & Streith, K.O. (2005). Examining user acceptance of computer technology: An empirical study of student teachers. *Journal of Computer Assisted Learning, 21*(6), 387–395.

Marriott, N, Marriott, P., & Selwyn N. (2004). Accounting undergraduates' changing use of ICT and their views on using the Internet in higher education. *Accounting Education, 13,* 117–130.

McCoy, S., Galletta, D., & King, W. (2007). Applying TAM across cultures: The need for caution. *European Journal of Information System, 16,* 81–90.

Moon, J., & Kim, Y. (2001). Extending the TAM for a world-wide-web context. *Information & Management, 38*(4), 217–230.

Myers, J. M., & Halpin, R. (2002). Teachers' attitudes and use of multimedia technology in the classroom: Constructivist-based professional development training for school districts. *Journal of Computing in Teacher Education, 18*(4), 133–140.

Ngai, E. W. T., Poon, J. K. L., & Chan, Y. H. C. (2007). Empirical examination of the adoption of WebCT using TAM. *Computers and Education, 48*, 250–267.

Phillips, L. A., Calantone, R., &. Lee, M. T. (1994). International technology adoption: Behavior structure, demand certainty and culture. *Journal of Business & Industrial Marketing, 9*(2), 16–28.

Reffell, P., & Whitworth, A. (2002). Information fluency. *New Library World, 103*, 427–435.

Riemenschneider, C. K., Harrison, D. A., & Mykytn, P. P. Jr. (2003). Understanding IT adoption decisions in small business: Integrating current theories. *Information & Management, 40*(4), 269–285.

Rotsaka, I. (2008). *Elementary school teachers' beliefs, attitudes and intentions concerning the educational use of simulations.* Unpublished Master's Thesis, University of Cyprus, Nicosia, Cyprus.

Sivo, S. & Pan, C. (2005). Undergraduate engineering and psychology students' use of a course management system: A factorial invariance study of user's characteristics and attitudes. *Journal of Technology Studies, 31*(2), 94–103.

Smarkola, C. (2007). Technology acceptance predictors among student teachers and experienced classroom teachers. *Journal of Educational Computing Research, 31*(1), 65–82.

Smith, B., Caputi, P., & Rawstorne P. (2000). Differentiating computer experience and attitudes toward computers: an empirical investigation. *Computers in Human Behavior, 16*, 59–81.

Sutton, S. (1998). Predicting and explaining intentions and behavior: Howwell are we doing? *Journal of Applied Social Psychology, 28*, 1317–1338.

Taylor, S., & Todd, P. A. (1995). Understanding information technology usage: A test of competing models. *Information Systems Research, 6*(2), 144–176.

Teo, T. (2006). Attitudes toward computers: A study of post-secondary students in Singapore. *Interactive Learning Environments, 14*(1), 17–24.

Teo, T., Lee, C. B., & Chai, C. S. (2008). Understanding pre-service teachers' computer attitudes: Applying and extending the Technology Acceptance Model (TAM). *Journal of Computer Assisted Learning, 24*(2), 128–143.

Teo, T., Su Luan, W., & Sing, C. C. (2008). A cross-cultural examination of the intention to use technology between Singaporean and Malaysian pre-service teachers: An application of the Technology Acceptance Model (TAM). *Educational Technology & Society,* 11, 265–280.

Tobin, K., Tippins, D.J., & Gallard, A.J. (1994). Research on instructional strategies for teaching science. In D.L. Gabel (Ed.), *Handbook of research on science teaching and learning* (pp. 45–93). New York: Macmillan.

Triona, L., & Klahr, D. (2003). Point and click or grab and heft: Comparing the influence of physical and virtual instructional materials on elementary school students' ability to design experiments. *Cognition and Instruction, 21*, 149–173.

Van der Meij, J. & de Jong, T. (2006). Supporting students' learning with multiple representations in a dynamic simulation-based learning environment. *Learning and Instruction, 16*, 199–212.

Venkatesh, V. (1999). Creation of favourable user perceptions: Exploring the role of intrinsic motivation. *MIS Quarterly, 23*(2), 239–260.

Venkatesh, V. & Davis, F.D. (1996). A model of the antecedents of perceived ease of use: Development and test. *Decision Sciences, 27*(3), 451–481.

Venkatesh V. & Davis F.D. (2000). A theoretical extension of the technology acceptance model: four longitudinal studies. *Management Science, 46,* 186–204.

Venkatesh, V., Morris, M.G., Davis, G.B., & Davis, F.D. (2003). User Acceptance of Information Technology: Toward a Unified View. *MIS Quarterly, 27,* 425-478.

Violato, C., Mariniz, A., & Hunter, W. (1989). A confirmatory analysis of a four-factor model of attitudes toward computers: A study of pre-service teachers. *Journal of Research on Computers in Education, 21,* 199–213.

Watson, D.M. (1993). *IMPACT – An evaluation of the IMPACT of the information technology on children's achievements in primary and secondary schools.* London: King's College.

Weinburgh, M. & Engelhard, G. (1994). Gender, prior academic performance and beliefs as predictors of attitudes toward biology laboratory experiences. *School Science and Mathematics, 94,* 118–123.

Wozney, L., Venkatesh, V., & Abrami, P. C. (2006). Implementing computer technologies: Teachers' perceptions and practices. *Journal of Technology and Teacher Education, 14*(1), 173–207.

Yu, J., Ha, I., Choi, M., & Rho, J. (2005). Extending the TAM for t-commerce. *Information & Management, 42*(7), 965–976.

Yuen, A. H.K., & Ma, W. W.K. (2002). Gender differences in teacher computer acceptance. *Journal of Technology and Teacher Education, 10*(3), 365–382.

Yuen, A.H.K., & Ma, W.W.K. (2008). Exploring teacher acceptance of e-learning technology. *Asia-Pacific Journal of Teacher Education, 36,* 229–243.

Zacharia, Z.C. (2003). Beliefs, attitudes, and intentions of science teachers regarding the educational use of computer simulations and inquiry-based experiments in physics. *Journal of Research in Science Teaching, 40,* 792–823.

Zacharia, Z. C. (2007). Comparing and combining real and virtual experimentation: An effort to enhance students' conceptual understanding of electric circuits. *Journal of Computer Assisted Learning, 23,* 120–132.

Zacharia, Z. C. & Anderson, O. R. (2003). The effects of an interactive computer-based simulation prior to performing a laboratory inquiry-based experiment on students' conceptual understanding of physics. *American Journal of Physics, 71,* 618–629.

Zacharia, Z. C., & Constantinou, C. P. (2008). Comparing the influence of physical and virtual manipulatives in the context of the physics by inquiry curriculum: The case of undergraduate students' conceptual understanding of heat and temperature. *American Journal of Physics, 76,* 425–430.

Zacharia, Z. C., Olympiou, G., & Papaevripidou, M. (2008). Effects of experimenting with physical and virtual manipulatives on students' conceptual understanding in heat and temperature. *Journal of Research in Science Teaching, 45,* 1021–1035.

Zanna, M. P., & Rempel, J. K. (1988). Attitudes: a new look at an old concept. In D. BarTal & A. W. Kruglanski (Eds.), *The social psychology of knowledge* (pp. 315–334). Cambridge, UK: Cambridge University Press.

Zint, M. (2002). Comparing three attitude-behavior theories for predicting science teachers' intentions. *Journal of Research in Science Teaching, 39,* 819–844.

APPENDIX: THE STUDY'S QUESTIONNAIRE

Part 1: Background information

1. Gender:
   - ☐ Male
   - ☐ Female

2. Years of teaching experience:
   - ☐ 0–5 years
   - ☐ 6–10 years
   - ☐ 11–15 years
   - ☐ >15 years

3. Education:
   - ☐ BSc (bachelors degree)
   - ☐ MSc (master degree)
   - ☐ Phd (doctorate)
   - ☐ Other _____

4. School district:
   - ☐ Rural
   - ☐ Urban

5. I am teaching at:
   - ☐ 1st grade
   - ☐ 2nd grade
   - ☐ 3rd grade
   - ☐ 4th grade
   - ☐ 5th grade
   - ☐ 6th grade
   - ☐ Other _____

6. I have received prior training on computer usage:
   - ☐ Yes
   - ☐ No

7. Computer usage during teaching:
   - ☐ Yes
   - ☐ No

Read carefully the following definition of what a simulation is and answer the questions that follow.

A simulation is a virtual, interactive dynamic visualization of apparatus and material, provided through a computer, which enables its user to have a learning experience that involves a process in which students virtually manipulate/interact with these material and apparatus to observe and understand the natural or material world (see also Figure A.1).

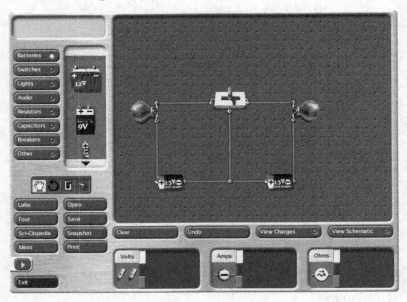

Figure A.1 Example of a simulation.

8. Have you ever use simulations during your teaching
   ☐ Yes
   ☐ No

   If your answer in question 8 was YES, then proceed to question 9. If your answer in question 8 was NO, then proceed to Part 2.

9. I have used simulations for teaching:
   ☐ Science
   ☐ Mathematics
   ☐ Language
   ☐ Music
   ☐ Geography
   ☐ History
   ☐ Technology
   ☐ Other _____

Parts 2 through 6

Please state below the degree you agree or disagree with the following statements. State your opinion by circling the number which characterizes you the most. Number **1** states complete **disagreement**, while number **5** states complete **agreement.**

Part 2: Perceived ease of use

	Complete disagreement				Complete agreement
Interacting with simulations is simple	1	2	3	4	5
A simulation is a tool that can easily be used	1	2	3	4	5
The use of simulations is simple	1	2	3	4	5
Learning to operate simulations is easy for me	1	2	3	4	5
It is easy for me to become skillful in using simulations	1	2	3	4	5

Part 3: Perceived teaching usefulness

	Complete disagreement				Complete agreement
Simulations for a fact can help me during my teaching	1	2	3	4	5
Simulations make my teaching easier	1	2	3	4	5
Simulations provide direct experiences with phenomena	1	2	3	4	5
Working with simulations makes teaching more interesting	1	2	3	4	5
Simulations relieve teachers of routine duties	1	2	3	4	5
Simulations may improve the overall quality teaching	1	2	3	4	5
Simulations could provide unique visualizations	1	2	3	4	5
Simulations make teaching more enjoyable	1	2	3	4	5
Simulations are a useful tool for teachers	1	2	3	4	5

Part 4: Perceived learning usefulness	Complete disagreement				Complete agreement
The use of simulations helps students acquire skills	1	2	3	4	5
Simulations enhance the understanding of complex concepts	1	2	3	4	5
The use of simulations enhances conceptual understanding	1	2	3	4	5
Simulations can support student group working	1	2	3	4	5
The use of simulations promotes the use of learning strategies (e.g., problem solving)	1	2	3	4	5
Simulations stimulate students' creativity	1	2	3	4	5
The use of simulations provides a better learning experience to the students than a traditional mode of instruction	1	2	3	4	5
The use of simulations helps students to give better explanations	1	2	3	4	5
Simulations actively engage students during a learning activity	1	2	3	4	5
The use of simulations increases interest towards learning	1	2	3	4	5
The use of simulations promotes the active participation of all students	1	2	3	4	5
The use of simulations gives the opportunity to manipulate all variables associated with the phenomenon under study	1	2	3	4	5
The use of simulations accommodates all students' different needs	1	2	3	4	5
Simulations make the presentation of all concepts (both real and reified) "observable" to the students	1	2	3	4	5
Simulations help students to self-regulate their own learning	1	2	3	4	5
Simulations improve students' achievement	1	2	3	4	5
The use of simulations promotes students' problem solving skill	1	2	3	4	5
The use of simulations develops students' critical thinking	1	2	3	4	5

Part 5: Attitude towards the educational use of simulations	Complete disagreement				Complete agreement
Simulations confuse me	1	2	3	4	5
I like using simulations in my teaching	1	2	3	4	5
I feel intimidated when I have to use simulations	1	2	3	4	5
The use of simulation is boring	1	2	3	4	5
The use of simulation disappoints me	1	2	3	4	5
The use of simulation wares me out	1	2	3	4	5
The use of simulation excites me	1	2	3	4	5
The use of simulation interest me little	1	2	3	4	5
Working with simulations makes my teaching enjoyable	1	2	3	4	5

Part 6: Behavioural intentions to the educational use of simulations	Complete disagreement				Complete agreement
I will try to learn anything that relates with simulations and their educational use	1	2	3	4	5
I would like to participate in seminars/courses that focus on the educational use of simulations	1	2	3	4	5
I would like to take simulation courses	1	2	3	4	5
I will pursuit to participate in courses of other teachers which use simulations during their teaching	1	2	3	4	5
I will like to be informed on recent developments in the educational use of simulations	1	2	3	4	5
I will teach courses that favor the use of simulations	1	2	3	4	5
I will use simulations as little as possible in my teaching	1	2	3	4	5
I will use simulations in my teaching, if I have the right equipment	1	2	3	4	5
I intend to use simulations in my teaching	1	2	3	4	5

CHAPTER 4

DEFENDING ATTITUDE SCALES

Per Kind and Patrick Barmby
Durham University

INTRODUCTION

The use of Likert items to create attitude scales has been under severe attack in science education research literature (Blalock, Lichtenstein, Owen, Pruski, Marshall, & Toepperwein, 2008; Gardner, 1996; Lederman, 2007; Munby, 1997; Osborne, Simon, & Collins, 2003; Reid, 2006; Schibeci, 1984). On one side, criticism has been directed towards the technical quality of attitude scales. Studies developing attitude measures tend to over-focus on simple indicators for internal consistency, such as Cronbach's Alpha, ignoring both test-retest reliability and validity (Gardner, 1996). Few studies, for example, use dimensionality analysis or test validity against external sources. They also disregard the problems of missing data (Blalock et al., 2008) and the use of ordinal data in continuous scales (Reid, 2006). On the other side, criticism has been directed towards the value of using attitude scales (Lederman, 2007; Osborne et al., 2003). This is related partly to the lack of technical quality as above, but more seriously, it is about the questionnaire approach being seen as not providing valuable data. It is, for example, claimed that many Likert-based questionnaires lack credibility because at-

Attitude Research in Science Education, pages 117–135

titude statements are made by researchers and not by students (Aikenhead & Ryan, 1992). In other words, they provide information about students' rating of the researchers' attitudes, rather than "true" attitudes held by students. A more compelling argument is that attitudes are far more complicated than what is revealed in Likert-based questionnaires. The questionnaires point towards subjects' expressed preferences and feelings for an object, but these are only "the tip of the iceberg" and do not necessarily relate to the behaviors that a pupil actually exhibits:

> [B]ehaviour may be influenced by the fact that attitudes other than the ones under consideration may be more strongly held; motivation to behave in another way may be stronger than the motivation associated with the expressed attitude; or, alternatively, the anticipated consequences of a specific behaviour may modify that behaviour so that it is inconsistent with the attitude held. (Osborne et al., 2003, p. 1054)

One possible reaction to these problems is to dismiss the use of "simple" attitude questionnaire. Another option is to look for new types of questionnaires that go deeper into the "iceberg," like the alternative presented in Bennet (2001). The line of argument presented in the current chapter, however, will take a different course from both of these and will set out to defend the use of Likert-based attitude scales. This argumentation is heavily based on the premise that attitude research serves many purposes, of which the simple attitude questionnaires may be useful for some but not for others. They are, for example, not suited for in-depth studies of the nature of students' attitudes and how these develop. This, of course, is important to know when trying to explain *why* students hold certain attitudes, but it is not the only information needed. Another purpose of attitude research is monitoring of students' attitudes. Both generally and in specific projects, it is interesting to know *if* students' attitudes change one way or the other, and to try to link such changes to other curriculum variables. In large-scale studies, such monitoring has to be based on simple instruments that can be easily administered. It is also a requirement that they are able to produce comparable results from one situation to another and over time when being used repeatedly. The argument in this chapter will be that Likert-based attitude questionnaires can be made to fulfill these requirements.

The supporting argument for attitude scales will partly be based on the view that techniques are available to create good scales. Evidence will be given through examples demonstrating a variety of techniques. As such, this chapter will draw on Rasch theory (Rasch, 1960) and will use suggestions by Smith (2004) and Bond (2003) for achieving validity in test instruments. The arguments presented, however, will also be based on a view that attitudes towards science, in some ways, are simple and "easy" to measure. No human trait or attribute, of course, can be measured in a straightforward way, but compared to other traits, attitudes have *some* characteristics that make them less complicated to measure. One of these is the fundamental link to *attitude objects*. Ajzen (2001) describes attitudes as evaluative judgements of attitudes objects. If we can decide what "object" we are looking for and what are the relevant emotional dimensions (good–bad, harmful–beneficial, pleasant–unpleasant, important–unimportant, etc.), we have made important steps towards deciding how to construct attitude measures. Importantly, we are then, as Ajzen and others researchers do, separating out the behavioral dimension of attitudes and not including this in the measure. Evidence that this works will be given in terms of examples from specific attempts to measure attitudes towards science. The overall point to be made is that attitudes may be measured meaningfully (which in measurement terms means with good validity and reliability) with simple means when scales are carefully constructed.

DATA AND METHODOLOGY

Empirical evidence for the arguments will be produced from an attitude questionnaire published in Kind, Jones, and Barmby (2007). When constructing the questionnaire, attempts were made to meet critiques within attitude research literature. Firstly, attitude objects were analyzed carefully, distinguishing between *science at school* and *science in society* (Gardner, 1996; Osborne et al., 2003), but also between other important nuances in students' attitudes towards science *within* both of these contexts. Relevant emotional dimensions were therefore implemented in judgmental statements towards objects, to produce items for attitude scales making sure that each scale focused, as far as we could see, on one particular object. Secondly, evidence that these scales exist as unique dimensions in the data was tested carefully in principle components factor analysis.

The questionnaire has subsequently been used in several studies, and data from two studies will be used in the current analysis. One study is the original survey for which the attitude questionnaire was developed (Kind et al., 2007). This was an intervention study with experimental and control groups, involving 932 pupils in a pre-test and 668 in a post-test. Cronbach's alpha for the attitude scales in pre- and post-testing is shown in Table 4.1, which also demonstrates the attitude objects and the number of items included in each scale. As can be seen, the scales had between five to eight items, and reliability values were mostly above 0.8. The other study is similar, but on a smaller scale. A single school carried out a science-in-society project using authentic problems taken from the local society in order to motivate science learning. One hundred and thirty-four students in one year group responded to the questionnaire in pre- and post-testing. The

TABLE 4.1 Cronbach's α Reliability Values for Each Attitude Measure

Measure	Cronbach's α	
	pre measure	post measure
Learning science in school (6 items)	0.89	0.92
Self-concept in science (7 items)	0.85	0.85
Practical work in science (8 items)	0.85	0.89
Science outside of school (6 items)	0.88	0.87
Future participation in science (5 items)	0.86	0.88
Importance of science and technology (5 items)	0.77	0.72
General attitude towards school (8 items)	0.85	0.85

reason for using this survey is that the questionnaire was slightly modified from the first survey. These modifications will be a focus in the argument of this chapter. To make clear which version of the test is being commented on, reference will made to the *original* and the *modified* questionnaire. Both studies were carried out in schools in the northeast of England.

Previous analyses (Kind et al., 2007) of the attitude questionnaire were set in conventional "classical" test theory. As mentioned above, in the current analysis, we have used Rasch analysis, which is based on "modern" test theory (Crocker & Algina, 1986). The latter has appeared as a response to criticisms against adding up *raw scores* (i.e., the actual score values to items) when making scales in social and psychological research (Bond & Fox, 2001). Rasch (1960) suggested a probability model for dichotomously scored items producing a scale which is based on comparing students' abilities with item difficulties. On this scale, both items and students are given a *measure* (i.e., an estimated value) based on the probability of a student (with a certain ability) getting an item (with a certain difficulty) right. If items and students share the same value on the scale, then there is a 50% chance of the student getting the item right, and this probability increases or decreases based on the relative position between item and student. An important feature of the scale is that the difference between item difficulty and student ability is theoretically invariant; if an item is one unit above a high scoring student and another item is one unit above a low scoring student, these differences between item difficulties and student abilities are the same. This makes the Rasch scale an additive conjoint measure (Luce & Tukey, 1964), which is not a guarantee in raw-score scales. Andrich (2007) has adapted the Rasch model to apply to Likert-based questionnaires. When this model is used in the current study, student *ability* refers to attitudes towards science (separated in scales for different attitude objects), and item *difficulty* to the level of endorsement to an item. The "easier" an item is, the more it is endorsed by students.

Using the Rasch approach offers solutions to two problems mentioned earlier. Firstly, it makes it possible to have continuous scales produced from ordinal data, since item difficulty (endorsement level) is taken into consideration when constructing the scale. Next, it has a way of handling missing data, since an estimated value can be produced on an item knowing the students' ability and the item's difficulty. It should be noted, however, that raw-data scales and estimated Rasch scales have high correlation. In the current study, the two types of scales on the same data have correlation values between 0.98–0.99, indicating that the problems caused by adding raw scores and by missing data are minimal for the studies considered.

Other benefits of the Rasch analysis are the techniques for validity analysis. Smith (2004) uses Messick's (1989; 1995) validity review criteria produced by the Medical Outcomes Trust (MOT, 1995) to describe eight facets of validity that can be investigated in Rasch analysis:

- Content validity—evidence of content relevance, representativeness and technical quality
- Substantive validity—empirical evidence that intended processes have been engaged by respondents to the items
- Structural validity—evidence for the credibility of the scoring structure
- Generalizability—evidence of invariance across tasks, time, groups, and contexts
- External validity—evidence of correspondence between constructs as expected
- Consequential validity—evidence concerned with implications of score interpretation and potential consequences of test use
- Responsiveness validity—the capacity of an instrument to detect change
- Interpretability—the degree to which qualitative meaning can be assigned to quantitative measures

These criteria and the techniques suggested by Smith have been used when analyzing the attitude scales. To simplify the presentation, we will present findings in three broader categories. The first is called the *construct-scale dialog* and looks at the alignment between theoretical understanding of the "construct" involved and the items in the scale. This includes, in particular, the content and substantive validity from the list above. The second is *structural evidence*, and includes the three following elements in the list: structural validity, generalizability, and external validity. The third is *outcome-related evidence* and includes consequential validity and responsiveness. The focus is then on the findings when using the questionnaire. More specific presentation of the validity criteria and the techniques used will be included in each section. In the Rasch-based analysis, we used the software Winsteps (Linacre & Wright, 2001). Other statistical analysis of data was carried out in SPSS.

CONSTRUCT—SCALE DIALOG

Bond (2003) describes the development of measurement scales as a "theory–practice dialectic." The understanding of a theoretical construct is seldom complete when producing a measurement scale, and the development process itself may add further information. Wiliam (2010), however, points towards a danger if the dialectic is too much on the side of the empirical data. The main focus, he claims, should be on the theoretical construct rather than the data. It is more important to ask *what we are measuring* rather than, from technical issues, *how well we are measuring*. In Rasch analysis we have important ways of conceptually examining the theoretical element in the scale, looking for evidence for content and substantive validity.

One way of doing this is by analyzing the ordering of items and persons in the scale. Figure 4.1 presents this ordering in an item-person map for the scale *Importance of Science and Technology* with data taken from the original measure. The scale is shown as a vertical line with students (#) to the left and items to the right. Each "#" represents 9 students of about the same attitude score. The values of the scale are logit units, shown to the far left. Zero is at the mean for items which are shown to the right of the line. The values plotted are positions between two categories in the Likert items, given by the title of the upper category. For example, "7d.5" indicates the point on the scale where students are at the boundary between the categories 4 and 5 on item 7d, and would have a 50% chance of being rated 4 or below and a 50% chance of being rated 5 or above. The scale had six items (7a to 7f), and from Figure 4.1, we can see their ordering from the most to the least difficult:

7d) Science and technology are helping the poor
7f) Scientists have exciting jobs
7c) The benefits of science are greater than the harmful effects
7a) Science and technology is important for society
7b) Science and technology makes our lives easier and more comfortable
7e) There are many exciting things happening in science and technology

7a, 7b, and 7e have almost identical difficulty and are therefore marked on the same position on the scale. The construct-scale dialog allows us to be able to explain this ordering of the items. For example, it seems reasonable that 7d has higher difficulty than 7c, since science may be beneficial

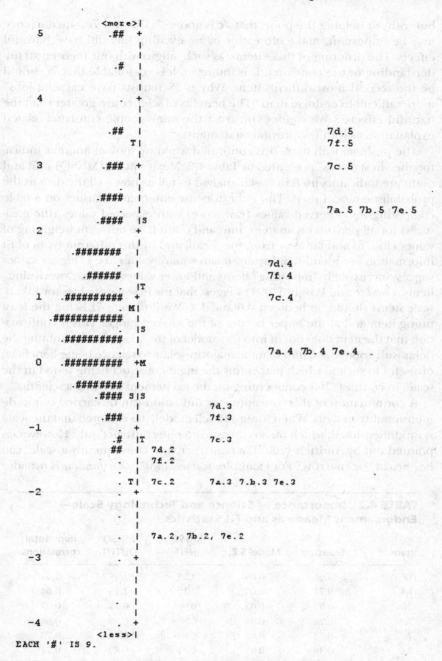

```
         <more>|
     5      .## +
               |
               |
             .# |
               |
     4         .+
               |
               |
             .## |                              7d.5
             T |                                7f.5
     3       .### +                             7c.5
               |
               |
             .#### |
             .#### |S                           7a.5  7b.5  7e.5
     2          .+
               |
        .######## |
               |                                7d.4
          .###### |                             7f.4
             |T
     1  .######### +                            7c.4
        .##########.
               |S
          .######### |
             .      |                           7a.4  7b.4  7e.4
     0    .######### +M
        .######## |
           .#### S|S
             .### |                             7d.3
               |                                7f.3
    -1         .# |T                            7c.3
             ## |                               7d.2
               |                                7f.2
               |
             . T|                               7c.2        7a.3  7.b.3  7e.3
    -2          .+
             .   |
               |
               |                                7a.2, 7b.2, 7e.2
               |
    -3          .+
               |
               |
             .   |
               |
    -4          .+
         <less>|
  EACH '#' IS 9.
```

Figure 4.1 Item-person map for *Importance of science and technology* scale. M is the mean, S and T indicates one and two standard deviations from the mean.

but without helping the poor; that 7c is above 7a, 7b and 7e, since science may be important, make life easier or be exciting but still have harmful effects. The ordering of these items, as such, aligns with our theoretical understanding of the construct. It is, however, less reasonable that 7f should be the second most difficult item. Why is "Scientists have exciting jobs" more difficult to endorse than "The benefits of science are greater than the harmful effects"? We cannot construct the same simple construct-related explanation as with the previous statements.

The problem with item 7f is confirmed when we look at another indicator, the "fit statistics" presented in Table 4.2. Mean square (MNSQ) infit and outfit are indicators used in Rasch analysis to tell us how well the data fit the probabilistic model used. The indicators use chi-square statistics on a table comparing the observed values (raw score) with expected values (the measures) for all persons on an item. Infit and outfit have different weighting of values close to and far away from the mean, and by this different types of fit information are identified. Ideally, mean square values are 1; higher values suggest increasingly "misfitting" items and lower values suggest "overfitting" items. Linacre and Wright (1994) suggest that mean square values for Likert-scale items should be between 0.6 and 1.4. We find item 7f being the least fitting item and at the upper border of the aspired range. This is confirmation that the item does not fit into the model of the construct. Examining the items with good fit, we find that the majority relate to science being beneficial or useful to society, which makes this the main construct being tested in the scale. In contrast, 7f is commenting on the excitement of being a scientist.

A continuation of this conceptual examination can be carried out in dimensionality analysis. When using a Rasch model, it is assumed that the scale is unidimensional, which means measuring one construct only. However, as pointed out by Andrich (2007), a construct, and consequently a scale, can be "broad" or "narrow." For example, *mathematical understanding* is broader

TABLE 4.2 Importance of Science and Technology Scale—Endorsement Measures and Fit Statistics

Item	Measure	Model S.E.	MNSQ INFIT	MNSQ OUTFIT	Item–total correlations
7f	0.60	0.05	1.35	1.38	0.66
7d	0.74	0.05	1.12	1.15	0.64
7b	−0.56	0.05	0.96	0.94	0.69
7c	0.24	0.05	0.85	0.89	0.68
7e	−0.53	0.05	0.88	0.86	0.73
7a	−0.48	0.05	0.81	0.82	0.71
Mean (SD)	2.82 (0.21)	0.02 (0.00)	0.99 (0.19)	1.00 (0.20)	

than *multiplication skills*, but they can both be seen as unidimensional constructs in the sense that they represent the main common underlying factor in a scale. When constructing a scale with good construct-scale alignment, we should understand this broadness and possible tensions between subconstructs involved. We should also be aware of *construct-irrelevant variance* (Messick, 1995) when there are factors other than those intended to be measured influencing the scale. A tool for carrying out such an analysis is Principal Components Analysis (PCA) on residual values between expected and actual values for the data. A correlation matrix is firstly produced for the differences between the raw score on an item and the value predicted by the Rasch model. If the construct identified in the model is the only factor in the scale, this correlation matrix should be "random." PCA is used to test this hypothesis; if it finds any factors among the residuals, this falsifies the unidimensionality of the scale. In practical terms, of course, this is not a dichotomous issue. Factors may be identified as more or less disturbing to the scale depending on strength (i.e., eigenvalues and factor loadings) and the qualitative nature of these factors (e.g., positively and negatively phrased items can be placed in separate factors). We therefore need to identify what the factors are and thereafter evaluate their threats.

In a closer examination of the scale given above, we identified the variance in the raw data (Table 4.3a), which may be compared with the variance in the modeled data. This gives another indication of how well the data fit the model. Next, PCA is carried out to identify factors in the residuals. The factors are presented as "contrasts" in Table 4.3b, and we see the number of factors identified and their strength in both eigenvalue units and percent. Factors with eigenvalues less than 1.6 have been found to not cause dimensionality problems in scales (Smith & Miao, 1994). The first factor only, therefore, is in need of further investigation. Table 4.3c presents the loading of each item onto this factor. Presenting the factor as a contrast is made clear by looking at items having opposite loadings (above and below the horizontal line). This reveals that items 7f and 7e are different from items 7a to 7d, meaning that not only the item "scientists having exiting jobs" (as found before) is threatening the unidimensionality of the scale, but that the same applies to the statement "There are many exciting things happening in science and technology." A narrower scale, in conceptual terms, could therefore be produced by focusing on "usefulness" and "importance" of science, excluding "excitement" entirely. This aligns with being more careful when choosing what emotional dimensions should be combined within the same attitude scale.

Summing up, the conceptual analysis gives strong evidence in terms of aligning the attitude scale with theoretical understanding. Data confirm expectation of development from "easy" to "difficult" as in Figure 4.1, about items fitting and not fitting the construct in Table 4.2, and further contrasts in

TABLE 4.3A   Variance in the Scale *Importance of Science and Technology*

Variance	Empirical		Modeled
	Eigenvalue	Percent	
Total variance	11.7	100.0	100.0
Raw variance explained by measures	5.7	48.7	49.3
Raw unexplained variance	6.0	51.3	50.7

TABLE 4.3B   Contrasts in Unexplained Variance

Unexplained variance	Eigenvalue	Percent of total variance	Percent of unexplained variance
Total	6.0	51.3	100.0
1st contrast	1.6	13.9	27.0
2nd contrast	1.5	12.6	24.5
3rd contrast	1.1	9.6	18.6
4th contrast	0.9	8.0	15.5
5th contrast	0.9	7.4	14.4

TABLE 4.3C   Items Loading onto the Factor Causing the 1st Contrast

Item	Measure	Loading
7f	0.60	0.77
7e	−0.53	0.52
7d	0.74	−0.05
7c	0.24	−0.40
7a	−0.48	−0.43
7b	−0.56	−0.64

the dimensionality in Table 4.3. Finding two items that *do not* fit the construct strengthens rather than weakens our understanding of the scale. An important point to be made is that the knowledge achieved through the analysis gives a basis to improve the scale. One obvious issue is removing the misfitting items, but Figure 4.1 also indicates a need for more difficult items, since the mean for items is lower than for persons, and we do not find any items matching the ability of the highest scoring students. As a result, a new scale was made for the modified questionnaire, taking this information into consideration:

7a) Science and technology make our lives easier and more comfortable

7b) The benefits of science are greater than the harmful effects

7c) We depend on science and technology to solve environmental problems
7d) Science and technology are helping the poor
7e) Society would be better with less influence from science

The improved statistics on this scale will be demonstrated below.

STRUCTURAL EVIDENCE

With the next set of evidence, we look to show that the scale is structurally consistent. In classical test theory, consistency is most commonly related to correlations between items in the scale, such as Cronbach's α or KR-20, but Smith (2004) suggests that a much wider set of indicators from the Rasch analysis should be taken into consideration. We will look at two different issues.

The first issue is respondents' use of the five categories in the Likert-scale. 'Consistent use' would mean all categories being used in the order as expected. Following up the example in Figure 4.1, we find a crude picture just by noticing that moving from a lower to a higher category (e.g., from 1 to 2) always increases the difficulty (because of the Rasch model being used, the spaces between the categories is identical for all items). Figure 4.2, however, gives a much closer inspection of the category use in prob-

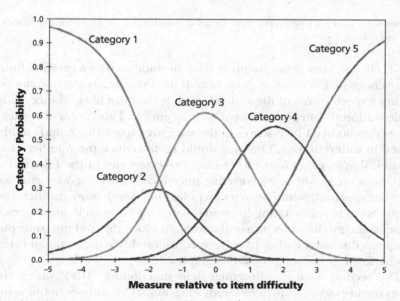

Figure 4.2 Category probability for items in original scale *Importance of science and technology.*

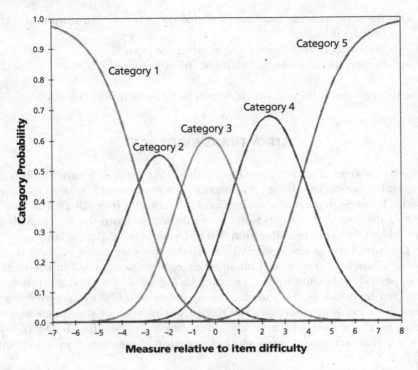

Figure 4.3 Category probability for items in modified scale *Importance of science and technology.*

ability curves. They show the probability of students with a certain attitude score choosing the category. A consistent use then means any one category having a specific area of the scale where it is the most likely choice. This is achieved for all categories except for category 2. This problem relates to an overlap between item scores in the scale area from 1 to 2, and could be solved in different ways. One way would be to reduce the number of categories; that is, using four rather than five categories in the Likert items. This, however, would be overreacting since mostly the categories work well. Another approach would be to try including items with more distinct scores on the attitude scales, trying to avoid the overlap between items. Figure 4.3 presents the modified version of the scale, in which this and the conceptual problems discussed earlier have been considered. We then find all categories having appropriate probability graphs.

The second issue is differential item functioning (DIF), which compares the use of the scale in different subgroups. Consistency in this setting means that an item should not be biased towards any subgroups, such as boys versus girls, or high-scoring versus low-scoring respondents. Figure 4.4

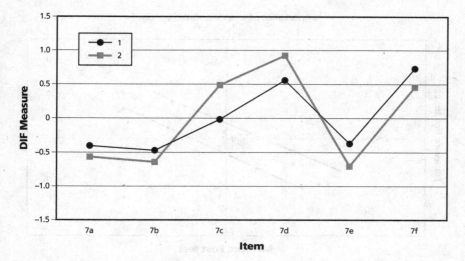

Figure 4.4 DIF between low scoring (1 = upper 50%) and high scoring (2 = lower 50%) students.

presents DIF measures for the upper and lower half of the pupils on the *Importance of science and technology* scale. The values in the graph are item difficulty for each subgroup, and Winsteps presents a test to control if these differences are statistically significant. A rule of thumb is that difference between groups (and it is important to keep in mind that we mean one item's difference in difficulty) should be at least 0.5 logits and with a probability $p < 0.5$ to merit further investigation. It is also important to take into consideration the number of items in a scale. A single item with a DIF exceeding the recommended level in a scale with many items will less likely be a problem. In light of this, it can be concluded that there is no strong DIF problem in the scale. Items 7c and 7d are more "difficult" (i.e., harder to endorse) for high-scoring than for low-scoring pupils, but the difference is at the level of what can be tolerated. Similar findings were achieved for the scale when testing for DIF on gender and year groups (the test included school year groups 7, 8, and 9).

A similar test to DIF is comparing difficulty measures of items across pre- and post-tests (i.e., a type of test-retest consistency). As in the DIF test, evidence for consistency would be that each item had the same difficulty measure in the two tests. As can be seen in the plot in Figure 4.5, this is the case for the *Importance of science and technology* scale. The dotted line marks identical measures (item measures being the same in both tests) and the two parallel lines the uncertainty area based on 95% confidence. As we can see, all items are within the uncertainty area.

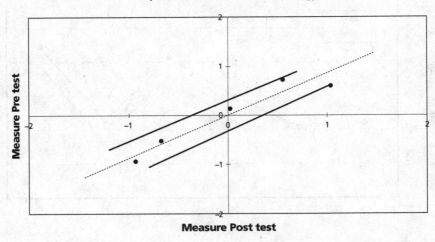

Figure 4.5 Plot of difficulty measures for items in the scale *Importance of science and technology*—Pre-test against post-test.

Two important points may be learned from this structural analysis. Firstly, it indicates that attitude scales can be produced with good consistency. In fact, when effort is made towards producing conceptually unidimensional scales, attitude measures exhibit better consistency. Secondly, it indicates that tools and procedures exist to identify and fix structural problems when they occur. The indicators used are only a few out of a series of tests that can be used in Rasch analysis to test for consistency.

USE OF THE ATTITUDE SCALE

Smith (2004) underscores the fact that responsiveness of a research instrument—that is, the capacity of an instrument to detect change—should be part of the validity evaluation. He therefore recommends comparing changes in scale scores before and after intervention. Two such issues will be presented as a last piece of evidence.

The first issue appeared as an unexpected finding in the study when the questionnaire was used for the first time. As mentioned previously, about 930 pupils from year groups 7, 8, and 9 were tested in a pre- and post-test, half of which took part in a given intervention (Barmby, Kind, & Jones, 2008). No significant difference was found between experimental and control groups, but another finding in this study was that of declining attitudes over the school years. This happened on average to all attitude constructs, but with important nuances. Boys, for example, did not have as negative a development in attitudes towards practical work as girls did, and the strongest negative development overall was on the scale *Learning science in school*. These findings were not surprising, but in themselves prove the sensitivity of the instrument to detect small nuances in pupils' attitudes. The surprising finding came when comparing results on pre and post tests alongside the overall development (over time) in year groups. As presented in Figure 4.6, the changes in attitudes from pre-test to post-test, which happened within a month, matched the trend found between the year groups. At first, we were puzzled about this finding, thinking that it may be a more general test-retest effect on the data. In light of other data, however, there is strong reason to believe that the attitude scales measured an actual development of attitudes within this short time period (about four weeks). For example, the trend was found for all attitude scales, in spite of these having different developments between one year group and the next. .

The second issue is from the intervention study with a single school, using the modified version of the attitude questionnaire. The intervention

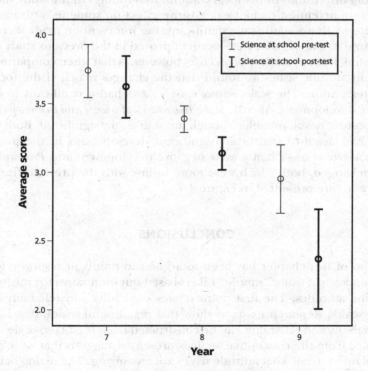

Figure 4.6 Attitude towards *Science at school*—Development in pre- and post-test scores for three year groups.

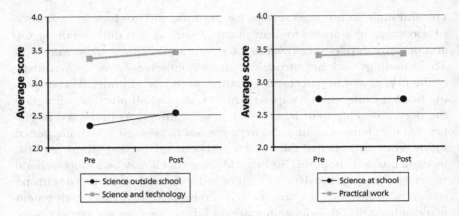

Figure 4.7 Attitude scores on pre- and post-tests in the intervention focusing on science in society.

focused on science in society and engaged students working on projects involving institutions in the local community. Findings in the study showed that the intervention had a weak positive effect on students' attitudes. Although not always statistically significant, the intervention made a change compared to the negative development proved in the previous study. Most interesting from a validity perspective, however, is that when comparing the different attitude scales we found that the changes matched the focus of the intervention. The scale *Science outside school* had a significant ($p < 0.5$) positive development. Also the scale *Importance of science and technology* had a weak positive development, although not statistically significant. Both these scales had negative statistically significant developments in the previous study. *Science at school* had a weak negative development and *Practical work* was unchanged, both which were more in line with the previous findings. The trends are presented in Figure 4.7.

CONCLUSIONS

The aim of this chapter has been to argue two points in response to the critiques against using "simple" Likert-based questionnaires for measuring scientific attitudes. The first point relates to validity and reliability of attitude scales. Research exists to show that meaningful scales, that is, with good validity and reliability, can be constructed from the Likert-scale items. Evidence from the conceptual analysis presented suggests that we are able to explain in detail what attitude scales are measuring. The dialog between construct and scale confirms that we should focus strictly on specific attitude objects and choose emotional dimensions accordingly. Any breach of this principle causes tensions in the scale and makes it more "blurred." It is, of course, possible to define *broader* attitude objects, following the terminology of Andrich (2007). Factor analysis presented by Kind and colleagues (2007), for example, suggested a combined attitude scale using scales looking at *Science at school, Science outside school* (e.g., visiting a science museum or watching science on TV), and *Future participation in science*. This scale is conceptually meaningful, as they are all science activities that children are potentially engaged in, and the scale proved technically consistent. *Narrow* attitude scales, however, are more sensitive and better at detecting changes. As shown in the data, narrow attitude scales are able to detect small nuances in students' attitude change, such as the falling attitudes occurring among pupils over a very short time period of a few weeks. Having instruments with such sensitivity is crucial when testing interventions and trying to find how they influence students' attitudes. The sensitivity of attitude scales, it seems, is a combination of conceptual focus and structural quality. Evidence indicates that both of these can be controlled, making us able to design good attitude scales.

The other point argued in this chapter has been about the techniques and procedures which make it possible to produce good attitude scales. In psychological and social measurements, there has been a major paradigm shift from classical to modern test theory (Bond & Fox, 2001). The new paradigm has a much stronger focus on validity and offers suitable techniques for such analysis. This chapter has demonstrated some of these techniques, but many more are available. These techniques, as underlined by Smith (2004), merge the issues of validity and reliability. They put forward a model that gives an expectation for what the data should look like. Testing consistency in the data means testing the validity of the model, because it is based on a theoretical understanding of the construct being tested. The dialog between scale and construct, between practice and theory, is therefore both a technical and an ideological issue. The techniques have proved to work already, but promise further development towards more valid and reliable attitude scales.

Bond (2003) argues for a next generation of scales in human sciences having the same quality as scales in the natural sciences. The scales should be "objective" in the sense that the units of the scale apply in the same way to any group in which the test is used. This, for example, would mean that we could test attitudes year by year in representative subsamples of students and examine if and exactly how attitudes are changing within a cohort. Objective comparison also would be possible across groups having been tested in different projects, even if the tests are not fully identical. Previously, this idea was impossible because no rationale or techniques existed to produce such instruments, but now this has changed. Modern test theory offers both ways of constructing scales and equating tests. In light of this, and the increased focus on validity, monitoring of scientific attitudes faces an entirely new situation and a brighter future. Denying the use of attitude scales would therefore be to throw out a very important baby with the bath water; this is a baby that is expected to make a change to future attitude research.

REFERENCES

Aikenhead, G., & Ryan, A. G. (1992). The development of a new instrument: Views on science-technology-society (VOSTS). *Science Education, 76,* 477–491.

Ajzen, I. (2001). Nature and operation of attitudes. *Annual Review of Psychology, 52,* 27–58.

Andrich, D. (2007). On the fractal dimension of a social measurement: I Report No. 3 ARC Linkage Grant LP0454080: Maintaining Invariant Scales in State, National and International Level Assessments. D Andrich and G Luo Chief Investigators: Graduate School of Education, The University of Western Australia.

Barmby, P., Kind, P. M., & Jones, K. (2008). Examining Changing Attitudes in Secondary School Science. *International Journal of Science Education, 30*(5), 1075–1093.

Bennet, J. (2001). The development and use of an instrument to assess students' attitude to the study of chemistry. *International Journal of Science Education, 23*(8), 833–845.

Blalock, C., Lichtenstein, M., Owen, S., Pruski, L., Marshall, C., & Toepperwein, M. (2008). In pursuit of validity: a comprehensive review of science attitude instruments 1935–2005. *International Journal of Science Education, 30*(7), 961–977.

Bond, T. G. (2003). Validity and assessment: a Rasch measurement perspective. *Metodología de las Ciencias del Comportamiento, 5*(2), 179–194.

Bond, T. G., & Fox, C. M. (2001). *Applying the Rasch model. Fundamental measurement in the human science.* Mahwah, NJ: Lawrence Erlbaum Associates.

Crocker, L., & Algina, J. (1986). *Introduction to classical & modern test theory.* Fort Worth, TX: Holt, Rinehart and Winston.

Gardner, P. L. (1996). The dimensionality of attitude scales: A widely misunderstood idea. *International Journal of Science Education, 18*(8), 913–919.

Kind, P. M., Jones, K., & Barmby, P. (2007). Developing attitudes towards science measures. *International Journal of Science Education, 29*(7), 871–893.

Lederman, N. G. (2007). Nature of Science: Past, Present and Future. In S. K. Abell & N. G. Lederman (Eds.), *Handbook of research on science education* (pp. 831–880). Mahwah, NJ: Lawrence Erlbaum.

Linacre, J. M., & Wright, B. D. (1994). Chi-square fit statistics. *Rasch Measurement Transactions, 8*(2).

Linacre, J. M., & Wright, B. D. (2001). *Winsteps* (Computer program). Chicago: MESA Press.

Luce, R. D., & Tukey, J. W. (1964). Simultaneous conjoint measurement: A new type of fundamental measurement. *Journal of Mathematical Psychology, 1*(1), 1–27.

Messick, S. (1989). Validity. In R. L. Linn (Ed.), *Educational measurement* (3rd ed.). New York: Macmillan.

Messick, S. (1995). Validity of psychological assessment: Validation of inferences from persons' responses and performances as scientific inquiry into score meaning. *American Psychologist, 50,* 741–749.

MOT, M. O. T. S. A. C. (1995). Instrument review criteria. *Medical Outcomes Trust Bulletin,* 1–4.

Munby, H. (1997). Issues of validity in science attitude measurement. *Journal of Research in Science Teaching, 34*(4), 337–341.

Osborne, J., Simon, S., & Collins, S. (2003). Attitude towards science: A review of the literature and its implication. *International Journal of Science Education, 25*(9), 1049–1079.

Rasch, G. (1960). Probabilistic models for some intelligence and achievement tests. Copenhagen: Danish Institute for Educational Research (Expanded edition, 1980, Chicago: University Chicago Press).

Reid, N. (2006). Thoughts on attitude measurement. *Research in Science & Technological Education, 24*(1), 3–27.

Schibeci, R. (1984). Attitudes to science: An update. *Studies in Science Education, 11,* 26–59.

Smith, J., E.V. (2004). Evidence for the reliability of measures and validity of measure interpretations: A Rasch measurement perspective. In J. Smith, E.V. & R. M. Smith (Eds.), *Introduction to Rasch measurement* (pp. 93–122). Maple Grove, MN: JAM Press.

Smith, R. M., & Miao, C. Y. (1994). Assessing unidimensionality for Rasch measurement. In M. Wilson (Ed.), *Objective measurement: Theory into practice* (Vol. 2, pp. 316–327). Norwood, NJ: Ablex.

Wiliam, D. (2010). What counts as evidence of educational achievement? The role of constructs in the pursuit of equity in assessment. *Review of Research in Education, 34,* 254–284.

CHAPTER 5

THE MULTIPLE RESPONSE MODEL FOR THE "VIEWS ON SCIENCE-TECHNOLOGY-SOCIETY" (VOSTS) INSTRUMENT

An Empirical Application in the Context of the Electronic Waste Issue

Yuqing Yu and Felicia Moore Mensah

INTRODUCTION

A major goal of science education today is to promote students' understanding of the nature of science, technology, and their interactions within society (American Association for the Advancement of Science [AAAS], 1990, 1993; National Research Council [NRC], 1996). To accomplish this goal, science educators have advocated the instruction of scientific knowledge in the context of Science-Technology-Society (STS) education. STS education supports students' natural tendency to integrate their personal understand-

Attitude Research in Science Education, pages 137–176

137

ings of the natural, technological, and social environments (Aikenhead, 1994; Layton, 1994; Ziman, 1994). Fundamentally, it is student-oriented; ultimately, it aims to promote students' social responsibility in collective decision making on issues related to science and technology (Aikenhead, 1994; Ziman, 1994).

Given the increasing importance of STS education, it is disconcerting to realize that few instruments are available to monitor students' views on STS topics. Additionally, the several standardized instruments that have been used over the decades have, by and large, been used with the erroneous assumption that students perceive and interpret the test statements in the same way as researchers do (Aikenhead & Ryan, 1992). For example, an empirical study showed that a number of students actually did not understand the meaning of the item "science knowledge is tentative," whenever students responded "agree" or "disagree" to the item (Aikenhead, 1979).

To diminish the problem of ambiguity (researchers assuming that there is no uncertainty between their interpretations and students' interpretation of the test statements), Aikenhead, Ryan, and Fleming (1989) developed the Views on Science-Technology-Society (VOSTS), a 114-item pool by employing empirically derived multiple-choice items, which are paraphrases of opinions expressed in interviews and/or written samples by respondents similar to those with whom the items are intended to be used. The VOSTS conceptual scheme includes the following topics: definitions of science and technology, the external sociology of science (mutual interactions with science, technology and society), the internal sociology of science (characteristics of scientists, social construction of scientific knowledge and technology), and the nature of scientific knowledge.

However, Aikenhead and colleagues' (1989) instrument is qualitative in nature and has two limitations. First, the item pool lacks a scoring system that allows researchers to use inferential statistical procedures. Second, respondents select just one choice, the one that better fits their opinion. Such a response model does not tell us anything about the non-selected options, except that they represent the respondents' opinion less than that which was chosen (Vazquez-Alonso & Manassero-Mas, 1999). Overall, the VOSTS instrument lacks a scoring system with which to apply inferential statistics and does not take advantage of the large volume of available information.

To address the limitations, a new model, the Multiple Response Model (MRM), has been proposed by Vazquez-Alonso and Manassero-Mas (1999). The primary goals of the MRM are to lend the VOSTS instrument itself to quantitative analysis, to maximize the use of available information, and to increase the precision of knowledge about respondents' attitude.

The purpose of this study was to empirically test the application of the MRM in the context of the electronic waste (e-waste) issue. The e-waste study explored the role of STS conceptions in adults' decision making about e-waste related issues. The choice of the VOSTS instrument as the tool to conduct the e-waste study is due to its inclusion of a broad range of STS topics. The VOSTS instrument allows the selection of relevant items to suit the study of a specific STS issue (Aikenhead & Ryan, 1992). The VOSTS instrument has been sensitive enough to detect differences in students who had studied an STS course and those who had not (Zoller, Ebenezer, Morely, Paras, Sandberg, West, et al., 1990). It also has been used successfully with university students (Fleming, 1988; Rubba, Bradford, & Harkness, 1996) and with high school teachers (Zoller et al., 1990).

The need for transforming a qualitative VOSTS response to a quantitative one was determined by the research interest of this study. The e-waste study sought to understand the relationship between STS conceptions and decision making—for example, how a student would likely take a particular action, given the student's responses to VOSTS items X, Y, and Z. Given the fact that other studies along the same direction have reported empirical challenges of qualitatively detecting an unambiguous relationship (Bell & Lederman, 2003; Zeidler, Walker, Ackett, & Simmons, 2002), a quantitative methodology that enables the meta-analysis of data is obviously compelling.

This study aimed to fill the void in the literature, since no empirical study has quantitatively assessed STS conceptions and sought quantitative relationships between STS understandings and socio-scientific decision making. Moreover, the MRM has not been tested in any empirical setting (Vazquez-Alonso & Manassero-Mas, 1999). This study is an initial step to provide an application of a theoretical developed model—the MRM—in an empirical setting. However, this study did not strive to examine the reliability and validity of the VOSTS instrument itself for the following reasons.

First, Aikenhead and Ryan (1992) argue that it is inappropriate to speak about the validity of an empirically developed instrument, such as the VOSTS, because the validity of empirically developed instruments is established by the "trustworthiness" of the method used to develop the items. Thus we assumed that the process used to develop the VOSTS items guarantees the trustworthiness of the instrument (Aikenhead & Ryan, 1992; Rubba et al., 1996). Second, Aikenhead and Ryan argue that the reliability of the VOSTS instrument is built upon the "dependability" of the results, which means that other readers should concur that, given the data collected by the researcher, the results make sense—that the results are dependable. As a result, since we agreed that the data presented by Aikenhead et al. (1989) make sense, we assumed the VOSTS instrument is reliable (Rubba et al., 1996). In other words, we assumed that the VOSTS instrument possesses an inherent validity that originated from the process used to develop it and the VOSTS instrument is reliable based on our agreement that the data presented by Aikenhead et al. (1989) make sense.

In summary, this study addressed the rationale to justify the MRM and demonstrated an empirical application of it by choosing several VOSTS items from the original VOSTS item pool. We assumed the validity of our selected VOSTS items because the MRM procedure does not result in any change of the method used to develop the VOSTS items. Nevertheless, we had to examine the reliability of our selected VOSTS item, because the MRM yields parametric data that fulfill the continuity and equal intervals of measures assumption that underlie parametric analysis procedures (Kerlinger, 1973). In other words, although the VOSTS instrument itself (without the MRM) is qualitative and reliable, the MRM generates parametric scores that require reliability assessment. Therefore, we examined the reliability of the selected VOSTS items, not the reliability of Aikenhead and colleagues' (1989) VOSTS instrument itself. In the following section, an overview of the MRM is provided.

OVERVIEW OF THE MULTIPLE RESPONSE MODEL (MRM)

A VOSTS item, on average, includes seven positions that express a domain of viewpoints about an STS topic. To obtain relational information between the options, the MRM attaches a rating scale to each position in a VOSTS item (Appendix A). The respondent assesses the fit between each position and his or her opinion on a scale of one to nine, with agreement increasing with the number. Each item then produces a set of scores that conforms to the respondent's general belief on the topic.

Next, the MRM assigns a value to each position, since the meaning of a respondent's rating scores depends on the character of each position. The MRM classifies positions into the following categories:

- A/Appropriate: The position expresses an appropriate view.
- P/Plausible: While not completely appropriate, the position expresses plausible points.
- N/Naïve: The position expresses a view that is inappropriate or not plausible.

The classification of positions across the three A/P/N categories provides the rational discrimination among positions and, therefore, distinguishes the most valuable views (the "appropriate" category) from the intermediate ones (the "plausible" category) and the valueless ones (the "naïve" category). Five expert judges, with respect to the current knowledge of history, epistemology, and sociology of science, assign all the positions.

Finally, the MRM involves the computation of an index for each item from respondents' direct rating scores. The computation depends on the

A/P/N character of each position. However, the computation is independent of the item's structure of choice (i.e., the number of positions per item) and the item's structure of categories (i.e., the number of "appropriate," "plausible," and "naïve" positions per item). Therefore, the computed index (in the range of -1 and +1) is an invariant parameter that is addable. An index close to -1 implies that the respondent holds a conflicting position with the expert panel, and hence, an improper attitude. By contrast, an index close to +1 implies that the respondent has a similar view with the expert panel, and hence, a proper attitude. The addable nature of the indices permits the application of the MRM with several VOSTS items simultaneously to obtain a unique parameter of a respondent's global belief toward a set of STS topics. A detailed explanation of the formulae and procedures to compute an item's index is displayed in Table 3 of Vazquez-Alonso and Manassero-Mas (1999, p. 242).

Vazquez-Alonso and Manassero-Mas (1999) argue that the MRM is not another version of a conventional Likert questionnaire, but that it is superior. According to these researchers, the scoring procedure of the MRM guarantees the unidimensionality for a set of items. They state that the MRM "defines a local system of meanings and weights that not only improve the efficiency of VOSTS pool . . . but also the multidimensional objection against attitudinal instruments" (p. 224).

This investigation applied the MRM in the context of the e-waste issue. It selected seven VOSTS items that are pertinent to the e-waste issue. The study addressed the following research question: How ought the Multiple Response Model (MRM) to be implemented to transfer qualitative STS responses to quantitative data? The next section discusses research methods of this study.

METHODS

This study employed a two-phased mixed methods paradigm to test the MRM application in the context of the e-waste issue. The convergence of quantitative and qualitative results ensured research quality and the trustworthiness of findings. At the first stage, a survey designed to assess respondents' views of STS topics and decision making about e-waste related issues was sent out (Appendix A). Those who had completed the survey formed the original sample of research participants. At the second stage, a subsample of the original sample was engaged in individual interviews and was explicitly asked to articulate their positions in response to several survey questions. This mixed methods approach is consistent with Rubba and colleagues' (1996) suggestion of not using the MRM procedure alone, as such a procedure is not a substitute for in-depth descriptive analysis of STS conceptions.

Survey Participants

This study focused on adults for three reasons. First, previous research indicates that K–12 students hardly possess the STS conceptions that most socio-scientific studies required (Aikenhead, 1987; Bady, 1979; Lederman, 1992; Miller, 1963; Rubba, Horner, & Smith, 1981). Second, the VOSTS instrument used in this study is not suitable for younger students such as 14-year-olds (Aikenhead & Ryan, 1992). Third, although children make some decisions about buying certain electronic devices, it is adults who generally make substantial personal and public decisions on how to deal with unwanted and obsolete electronic wastes.

Survey recipients were graduate students at a large urban university in the Northeastern United States. Students from the departments or programs of arts and humanities (A&H); human development (HD); computing, communication, and technology education (CCTE); mathematics education; and science education were contacted by a recruiting email. The email briefly introduced the study, included the link to an online e-waste survey, and was sent to 766 email addresses by administrators of relevant departments. To attract respondents, the first 25 respondents were promised to receive a small thank-you gift. The survey was kept open for 10 days and closed after no responses were received for three consecutive days.

The survey received 106 responses: 74 respondents completed it, and 32 respondents partially completed it. The 74 respondents who completed the survey formed the original sample of this study. The self-selection nature of the respondents inevitably resulted in sample bias. Presumably, the respondents responded to the survey invitation for the following reasons: (1) They had science and/or technology backgrounds and would be interested in a socio-scientific issue; (2) they were environmentally conscious persons and would be interested in an environmental problem; and (3) they understood the difficulty of data collection for a doctoral research study and were willing to help. The selection bias of this study limits the generalizability of research findings to other populations. The sample included 68% female and 32% male. The skewedness of gender distribution in the sample results from the intrinsic imbalance of female and male student composition of the institution (i.e., matriculated students of the institution are 77% women and 23% men).The gender asymmetry is inherent sample bias and can hardly be minimized. In terms of respondents' academic backgrounds, 24% were from A&H, 18% from HD, 24% from CCTE, 18% from mathematics education, and 16% from science education. Overall, the respondents with and without adequate science and technology backgrounds had comparable representativeness in the sample. Table 5.1 summarizes respondent composition.

TABLE 5.1 A Summary of Respondent Composition

Department/Program	Number of email addresses	Number of respondents	Percentage of respondents in the sample
A&H	434	18	24%
HD	89	13	18%
Mathematics Education	170	13	18%
CCTE	46	18	24%
Science Education	27	12	16%
Total	766	74	100%

Interview Participants

The sample for the interviews was solicited by a convenience sampling approach. Willingness to participate was the only factor that contributed to an individual's inclusion in the sample. Out of the original sample of 74, eight participants responded to the email invitation. The eight participants formed the interview sample and all of them received $20 cash on site. Five were females and three were males. Two participants were international students who come from Israel and Ukraine. One participant was Asian American; one was Hispanic American; and four were White Americans. All participants were in the age range of middle twenties to early thirties.

The academic background each of the participants had experienced is a notable feature of the sample. Three participants (P1, P6, & P8) were in the field of applied linguistics and would be English teachers soon. Participant 3 had a background in political science and had been a teacher who taught citizenship, justice, and globalization. Participant 7 just completed his first year in the CCTE program and was looking for a research topic for his thesis. Three participants (P2, P4, & P5) were in the cognitive psychology program, but each of them had different science and technology backgrounds: Participant 2 was teaching courses in the CCTE program and was conducting her doctoral dissertation research on psychology of media; Participant 4 had gotten a master's degree from the CCTE program and was now in the doctoral program in cognitive psychology; and Participant 5 was investigating inquiry skills in the context of science education and was pursuing a doctoral degree in cognitive psychology. Overall, the sample was comprised in a way that participants with and without adequate science and technology backgrounds were represented equally. Although exploring the role of content knowledge in socio-scientific decision making was not an explicit goal of this study, it is important to provide a "thick descrip-

TABLE 5.2 Demographic Information of the Interview Participants

	Gender	Nationality & Ethnic group	Degree program	Current department & program	Previous experiences
P1	F	U.S., Asian American	Master's	A&H: Applied linguistics	Applied linguistics
P2	F	U.S., White	Doctoral	HD: Cognitive psychology	Was teaching a course in the CCTE program
P3	M	Israel	Doctoral	A&H: Political science	A teacher of citizenship, justice, and globalization
P4	M	U.S., Hispanic	Doctoral	HD: Cognitive psychology	Awarded a master's from the CCTE program
P5	F	U.S., White	Doctoral	HD: Cognitive psychology	Was conducting research of inquiry skills in science education
P6	F	Ukraine	Master's	A&H: Applied linguistics	Applied linguistics
P7	M	U.S., White	Master's	MST: CCTE	Completed the first year in the CCTE program
P8	F	U.S., White	Master's	A&H: Applied linguistics	Applied linguistics

tion" of participants involved in a qualitative study (Guba & Lincoln, 1989). Table 5.2 summarizes interview subsample characteristic.

Data Collection from Survey Instruments

STS Conceptions

The items to test understandings about STS topics were chosen from the VOSTS item pool (Aikenhead et al., 1989). The seven selected items were from two major sections—Social Construction of Technology and Influence of Science/Technology on Society. They were 80211, technological imperative; 80131, technological decisions; 40161, concern and accountability for risk and pollution; 40212, technocratic versus democratic decisions; 40221, contribution to moral decisions; 40311, trade-offs between positive and negative consequences of science and technology; and 40451, technological fix to social and practical problems. We selected these items because they were most relevant to the e-waste issue. All the selected items express views from a customer's perspective and assess conceptions related to the nature of technology, pollution and moral and ethic issues. Table 5.3 lists topic statements and topic areas.

TABLE 5.3 Topic Statements and Topic Areas of the VOSTS Items

Item Number	Topic Statement	Topic Area
80211	Technological developments can be controlled by citizens.	Technological imperative
80131	When a new technology is developed, it may or may not be put into practice. The decision to use a new technology depends on whether the advantages to society outweigh the disadvantages to society.	Technological decisions
40161	Heavy industry has greatly polluted North America. Therefore, it is a responsible decision to move heavy industry to underdeveloped countries where pollution is not so widespread.	Concern and accountability for risk and pollution
40212	Scientists and engineers should be the ones to decide on North American air pollution standards because scientists and engineers are the people who know the facts best.	Technocratic versus democratic decisions
40221	Science and technology can help people make some moral decisions.	Contribution to moral decisions
40311	We always have to make trade-offs between the positive and negative effects of science and technology.	Trade-offs between positive and negative consequences of science and technology
40451	We have to be concerned about pollution problems that are unsolvable today. Science and technology cannot necessarily fix these problems in the future.	Technological fix to social and practical problems

E-waste Decision Making

Decision making in the study is referred to an individual's intention to participate in environmentally friendly actions, since no real behavior was actually observed. Six statements, adopted from Saphores, Nixon, Ogunseitan, and Shapiro's (2006) study, were used to measure respondents' willingness toward environmentally friendly action. Their instrument was adopted because of its reported reliability among items and an acceptable Cronbach's alpha (Cronbach's alpha larger than .7). More specifically, these statements asked the respondents to rate the degree of willingness they would devote to personal and public actions that are friendly to the environment. Intention statements were included in a five-point semantic-differential scale. A higher value demonstrates more willingness to support the environment (Appendix A for survey questions).

Interview Protocol

The interviews were conducted individually in a private reserved room in the library and lasted between 25 to 45 minutes. All interviews were audiotaped for transcription. Each interview began with a general introduction of the interviewer and the study, followed by watching a video—*E-waste: Dump on the Poor* (http://www.youtube.com/watch?v=ExzsqTFwV3Q&feature=related). Next, the participants were asked to elaborate their survey responses to the two VOSTS items that had demonstrated statistically significant relationships with decision making. More specifically, the participants were asked to clarify their views regarding the practice of moving polluting industries to underdeveloped countries as well as their beliefs about the relationship between moral decisions and science and technology. They were also asked to provide concrete examples about how science and technology help or do not help to make moral decisions in their life. Finally, the participants were asked about potential action they would take to protect the environment and to help other people become more aware of the e-waste issue (see Appendix B for interview questions).

Data Analysis

Analysis of the VOSTS Items

A panel of five expert judges, who are faculty members of the Science Education Program of the institution who represent the fields of science education and STS, was charged with independently classifying every position in each of the selected VOSTS items by the A/P/N scheme (Appendix C for the categorization of the VOSTS items). Agreement among at least three of the five panel members was sought in order to categorize a position as A, P, or N. The choices "I don't understand," "I don't know enough about this subject to make a choice," and "None of these choices fits my basic viewpoint" were not included in this study, because empirical evidence demonstrates that 90% of the time college students were able to describe their views by selecting one of the empirically developed answers (Rubba et al., 1996). When the panel's classification work resulted in a tie, we used discretion and intervened to assign the position to a category on the basis of the appropriateness of this position in the e-waste context. After categorization, computation of attitude indices was performed according to the procedures and formulae explained in Table 3 of Vazquez-Alonso and Manassero-Mas (1999, p. 242).

Relationships among the Constructs

Internal consistency test was performed to examine internal reliability among the constructs under investigation. Internal consistency examines whether every item in a scale shares common variance with at least some other items in the scale, but does not entail unidimensionality (Gardner, 1995). Cronbach's alpha is the statistic that is most widely used for estimating internal consistency. It is defined in terms of the ratio of the sum of the variances of the individual item scores to the variance of the scale score. Alpha is maximized when every item correlates well with at least some other items in the scale (Gardner, 1995). For studies for exploratory purposes, a moderate Cronbach's alpha of .60 is the lenient cut-off; for general research, an acceptable alpha should be within the range of .7 to .9.

Factor analysis, such as principle components analysis (PCA), was used to test whether a set of items all load on a common factor—that is, whether they all measure a unidimensional construct. If the unidimensionality assumption was not met, items were analyzed individually (Gardner, 1995). Finally, simple linear regression was employed to detect relationships between the VOSTS items and decision making. Given the diversity of constructs under investigation, a multivariate analytical approach would have been appropriate. However, with only one dependent variable, it is not possible to apply multivariate statistics (Yang & Anderson, 2003). Rather, ANOVA tables were constructed to present these relationships. A demonstration of a statistically significant F value at a significant level of less than .05 ($p < .05$) suggests the existence of a correlational relationship between two constructs.

Interview Data Analysis

The interview data were analyzed using the grounded theory method (Strauss & Corbin, 1998). The interview transcripts were read through several times while notes were taken to discover emergent patterns. To ensure quality, 25% of interview transcripts were taken back to the participants and were reviewed by them as a form of member checking (Guba & Lincoln, 1989). Next, we attempted to "saturate" each pattern, that is, to look for emergent themes that represent the pattern and continue to look until new information obtained did not provide new insight into the pattern. Finally, patterns and themes from the interviews were compared with the survey responses to seek convergence of results. The findings are reported in the section below.

RESULTS

Mean Scores of the VOSTS Items

Grand average indices (GAIs) are used to describe the descriptive statistics of the VOSTS items. GAI is defined as the average index score among all the respondents. GAIs, in the range of −1 to +1, imply whether average respondents hold views that are aligned with (close to +1) or conflicting with (close to −1) those of the expert panel. Table 5.4 displays the values of GAIs and their responding standard deviation.

TABLE 5.4 GAI, Grand Average Index, of Each VOSTS Item

Item Number	GAI1	Standard Deviation	Minimum	Maximum
40161	.385	.255	−1	1
40221	.355	.287	−1	1
80211	.284	.277	−1	1
40212	.276	.235	−1	1
40311	.213	.233	−1	1
40451	.197	.254	−1	1
80131	.181	.325	−1	1

Item 40161, which examines the risks associated with relocating polluting industries to poor countries, has the highest GAI. This result demonstrates that compared to other VOSTS items, the respondents agreed with the expert panel most on item 40161. The respondents believe that it is not responsible to move heavy industries to poor countries. It may well be that item 40161 is the topic with which the respondents were most familiar, as the topic has often been discussed in media and in school. Therefore, the respondents were able to express proper views.

Item 40221, which explores the role of science and technology in moral decisions, has the second highest GAI. The respondents generally agreed with the expert panel that science and technology can help people make moral decisions in the sense that science and technology can provide people information, which helps cope with the moral aspects of life. However, the respondents had disparate views on the argument that science and technology cannot help make moral decisions. Some respondents fully agreed with the argument, some felt neutral about it, and some fully disagreed with it. Varied views on the role of science and technology in moral decisions reflect respondents' progressive or conservative perceptions about the nature of science and technology.

The rest of the VOSTS items have low GAIs. It is probable that the respondents were relatively unfamiliar with these topics or had rarely thought about them, and hence, were not able to deliver sufficient views. Moreover, item 40161 and item 40221 are the only ones that were found to have statistically significant correlational relationships with decision making, although their effect sizes are moderate (i.e., item 40161 explains 14.9% of variance of decision making, and item 40221 explains 10.0% of variance of decision making—see Table 5.5). These two topics were asked again in the follow-up interviews so that the participants could articulate their positions. This is the place where quantitative and qualitative data are integrated and triangulated.

TABLE 5.5(A)   ANOVA Table for the Association Between Item 40161 and Decision Making

ANOVA[b]

Model	Sum of square	df	Mean square	F	Sig.
Regression	.149	1	.149	6.909	.010*[a]
Residual	1.554	72	.022		
Total	1.703	73			

Note: [a] Predictors: (Constant), Item 40161; [b] dependent variable: Decision making.
* $p < .05$.

TABLE 5.5(B)   ANOVA Table for the Association Between Item 40221 and Decision Making

ANOVA[b]

Model	Sum of square	df	Mean square	F	Sig.
Regression	.100	1	.100	4.512	.037*[a]
Residual	1.602	72	.022		
Total	1.703	73			

Note: [a] Predictors: (Constant), Item 40221; [b] dependent variable: Decision making.
* $p < .05$.

Internal Reliability and Dimensionality

Internal consistency test was performed to examine internal reliability among the set of VOSTS items. This study failed to identify a significant Cronbach's alpha value. Internal consistency test showed an unacceptably low Cronbach's alpha value of .344. This value is much lower than the cut-off value for an exploratory study. The subsequent PCA failed to improve the alpha to reach the cut-off value of .6. Moreover, the collective treatment of the VOSTS items as a single set did not improve the quality of analysis that looked for relationships between STS conceptions and decision making. Therefore, the item scores should not be added together to generate a single scale. Rather, the VOST items were analyzed individually, as these items tested different aspects of STS.

It is probable that selecting several VOSTS items from the same topic area, rather than selecting items from different topic areas as practiced in this study, might increase internal reliability. However, Vazquez-Alonso and Manassero-Mas (1999) highlight that the MRM is different from and superior to a traditional Likert-scale questionnaire in the sense that the MRM guarantees unidimensionality, and this is not dependent upon the number of test items selected from the pool. The researchers claim, "The individual consistence on every item of this construct [the construct that measures the attitudes' degree of appropriateness with respect to an adequate conception of the science] guarantees the unidimensionality for a set of items" (p. 244). These researchers do not require a set of items from the same topic area, either. They simply state that the MRM permits "a natural extension to applications with several VOSTS items simultaneously" (p. 243). Therefore, we did not worry about the assumption of unidimensionality when selecting VOSTS items. Nonetheless, the empirical results do not support the unidimensionality assumption of the MRM. This point is discussed in greater detail in the next section.

Interview Results

The two items that were significantly linked to decision making were asked again in the interviews: item 40161—Moving polluting industries to underdeveloped countries, and item 40221—The role of science and technology in moral decisions. Although the interviews were conducted eight months after the survey, none of the participants changed their views and all of the participants expressed consistent views with respect to their survey responses. The findings are reported below.

Item 40161: Moving Polluting Industries to Underdeveloped Countries

All participants agreed that it is not a responsible practice for developed countries to move polluting industries to underdeveloped countries. Developed countries should eliminate or reduce pollution in their own land and not export pollution to underdeveloped countries that do not have technologies and funding to solve it. Four patterns—economic considerations, concerns about global effects of pollution, call for responsibility and empathy—were identified from interview responses.

Economic Considerations

One major pattern that the participants displayed in the interviews was their recognition of the economic factors involved in the business of exporting polluting industries. The participants acknowledged on one hand that developed countries do not have the right to pollute undeveloped countries; they on the other hand emphasized that the business of exporting provides poor people jobs and a living. According to the participants, although developed countries are taking advantage of cheap labor that underdeveloped countries offer, underdeveloped countries cannot decline the business because so many people need jobs. The following quotes reflect this pattern:

Interviewer (I): The United States is moving heavy industries to underdeveloped countries. Do you think this is an acceptable practice?

Participant 1 (P1): I think we are definitely taking advantages of our role. We are taking advantage of the fact that there are so many people in China that need work. We like that, because it's cheap labor. But at the same time, China cannot refuse it, because they need the work. We are taking advantage of that. I don't think this is acceptable.

P6: This is definitely damaging to the underdeveloped countries. This is like part of the global tendency that moves everything hazardous or work to underdeveloped countries. The U.S. tries to get rid of everything bad. I think it's not acceptable. But I think it's also part of the system. It's bad for poor people, but that's how they live.

Concerns about Global Effects of Pollution

The second emergent pattern was concern about global effects of pollution. The participants emphasized that exporting pollution to other places is hiding the problem, not solving it. As the world is a united community, pollution in other places will eventually affect the United States. They highlighted that the earth is a balanced system that can be altered in deleterious ways by human activity that all too often is understood only as a local problem, if at all. One participant humorously commented that he wishes we could send wastes to outer space; however, he worried that we would then start Star Wars. The following quotes indicate that the participants worried about global effects of pollution:

P2: It seems that something has to be done now, because it is going to grow to the point that it involves the whole planet whether we ship to another country or not. We only have one planet.

P7: Shipping waste to poor countries is just an easy way to get rid of it. Although you don't see it [pollution], it doesn't mean that it won't come back to haunt you. I wish we could send it to the outer space. But then we'd start Star Wars.

Call for Responsibility

The third pattern was participants' call for taking responsibility. They articulated this pattern from two perspectives. On one hand, the United States should take responsibility to dispose of e-wastes in its own land. The participants had assumed that the United States, as one of the most advanced countries, should lead in action to solve pollution problems. When they learned that the United States is exporting toxic e-wastes to underdeveloped regions, they considered exporting very unacceptable and urged the United States to take responsibility to treat the problem by itself. On the other hand, the participants blamed the wasteful lifestyle that people of the United States have had and argued that American people should be responsible for their lifestyle. The following quotes are examples of the pattern of responsibility demonstrated in the interviews:

> **P2:** No, absolutely not. I don't think it's acceptable. I am surprised the U.S. is the leader doing that. Actually, I don't think that's [the action of exporting] really well brought out in media. I am surprised that we are not in that action taken. We should treat our problem by ourselves.
>
> **P8:** I just think the U.S. should take more responsibility for the very wasteful lifestyle we have. We think that if we send things away they will disappear. But that's not true. It's clearly affecting the life of people in other countries and will ultimately affect our own life.

Empathy

One participant demonstrated strong empathy toward poor people and called for the rest of the world to protect those who themselves have the least ability to change. She highlighted that manufacturers should develop advanced technologies to reduce or eliminate toxic chemicals so as to protect poor people who sacrifice their health to work. Although only one participant explicitly displayed empathy regarding this interview question, the participants generally demonstrated emotions and empathy when responding to other interview questions. The following is her quote:

> **P5:** Some people have to take whatever the job they can get. And it might be damaging their health. So it's up to the rest of the world to make the moral decision so that when the products become waste, we don't want them to be damaging to people or the environment.

Overall, the participants agreed that moving polluting industries to underdeveloped countries is an irresponsible action and revealed the patterns of economic considerations, concerns about global effects of pollution, call for responsibility, and empathy.

Item 40221: The Role of Science and Technology in Moral Decisions

The second interview question was about the role of science and technology in moral decisions. When asked about the question, two participants requested a clarification of the meaning of "moral decisions" in this context. One participant commented that he had never realized that science and moral decisions were related. However, this is the same question asked in the survey. It seemed that not every respondent understood survey questions equally well, even if they still made responses to survey items. Three patterns—benefits of science, concerns about science, and the supporting role of science in moral decisions—were emergent from the interviews.

Benefits of Science

The participants agreed that science and technology provide information that would help them to make informed decisions. In particular, they argued that the Internet and online search engines make it easier for the general public to acquire necessary information. The participants recognized the benefits that science and technology bring. The following quotes reflect their acknowledgement of benefits:

> **I:** Do you believe that science and technology can help people make moral decisions? If so, in what specific ways?
>
> **P2:** I just guess it's an easier way to gather information. But I believe persons have to make their own decisions. But now it's in such a digital age and you can type anything in the search engine, and you have more opinions and more availability that you would decades ago. So, I think so.
>
> **P6:** Yes, of course they can. I think that there are some discoveries in science that scientists make about certain areas of life and scientists educate the public about those problems. They publish articles in popular magazines, go to TV, try to talk, and get people concerned about these problems. Of course, these in the end will influence moral decisions as well, because people will know more and they will think more about this.

Concerns about Science

In addition to benefits, the participants also expressed concerns about potential moral challenges that science and technology can cause. The pattern has two themes—conflict with traditional belief systems and unsustainable application of technology. One participant pinpointed that science can sometimes conflict with traditional belief systems. For example, science can prolong one's life, but can also make the person to die unnaturally and painfully under a very ill condition. Therefore, science may conflict with one's belief to live naturally and die naturally, and consequently, result in a moral dilemma. Another participant underscored that the way that technology is currently used is not sustainable: Technology is used to produce for more consumption and disposal. Moreover, technology itself demands moral concerns, because we were not terribly concerned, at least, at the individual level about the impact of many pollution problems when we did not have the technology. The following quotes present the participants' concerns:

> **P4:** Woo, that's a hard question. I do believe, yes, definitely. Science and technology can help people make moral decisions. But I think that there is potential for both [good and bad moral decisions] when it comes to science and technology and morality. There is a potential for good things or bad things. Science can conflict with traditional belief systems and cause moral dilemmas. And regardless, I think yes, science and technology can give people opportunities to make moral decisions. I do think so.
>
> **P7:** Yes, I think that information that comes out of science especially can help us make decisions to plan how we'd use technology in responsible ways. I don't think that the way technology currently produces furthers sustainability. It is more produced for consumption and disposal. I know that science seems to be probably informing that it should be more sustainable. But technology we are getting isn't following.

The Supporting Role of Science in Moral Decisions

Finally, the participants stressed the supporting role of science and technology in moral decisions. According to these participants, individual values and beliefs ultimately determine how a person would use the information provided by science and technology. The following are quotes discussing the supporting role of science and technology in moral decisions:

> **P5:** I think, yes, absolutely, because I feel that more information we have about anything, more tools we have to make a

moral decision. I don't think science tells us what is moral, but science tells us the facts. And we can use our own moral competence to value these facts.

P8: Possibly. I mean it can help people have access to more information, like the Internet. So I think more information usually helps people make well informed decisions. But I don't know science and technology alone can help people make moral decisions.

P3: I try to answer. Stop me if I go too much. As I said, I come from social studies and I like philosophy. So I don't know much about science. I think when you talk about moral issues, it's a philosophical question. What happens when you bring in the science aspect? It gives some kind of objectivity to it. Some think that scientists can say what they want and they are not biased by any philosophical point of view. I don't think that's true. Even scientists are influenced by their own bias, by their own philosophical point of view. That's ok. So the technology is only a tool, built on the philosophical base.

Moreover, the participants were asked to provide concrete examples of how science and technology help or do not help make moral decisions. Most participants gave general examples such as decision making about issues related to global warming and green houses. Two participants cited examples that were based on their personal experiences. For example, Participant 5 said that learning about a fetus in utero with her doctor let her determine that she would, under no circumstances, have an abortion. Participant 4 explained the supporting role of science in his grandma's case when she suffered severely from a cancer. Although the advance of science and technology provides her the opportunity to live with the aid of a machine in her body, she refused to use it and chose to die naturally. The participant highlighted that it was not scientific facts and discoveries but his grandma's personal values that finally determined her decision.

Overall, the participants expressed three patterns when considering the relationship between moral decisions and science and technology—benefits of science, concerns about science, and the supporting role of science. In addition, most participants provided non-personal examples regarding the role of science in their moral decisions. This is probably because the participants rarely contemplate the role of science when they make moral decisions in real life.

Triangulation

Moving Polluting Industries to Underdeveloped Countries versus
Item 40161

Interview data are integrated with survey responses to assess the trust-worthiness of quantitative and qualitative data. Table 5.6 summarizes each participant's rating on every statement of item 40161, which examines the risks associated with relocating polluting industries to poor countries. The participants rated the agreement between each statement and his or her opinion on a scale of one to nine, with agreement increasing with the number. According to Table 5.6, all participants disagreed that heavy industry should be moved to underdeveloped countries (statement a), as they all gave a very low score on this statement. Likewise, the participants articulated the same position in the interviews. Therefore, quantitative data matches qualitative data.

Moreover, all participants acknowledged the economic factor inherent in the exporting business, as all of them gave a high score on statement b. Comparably, the participants explicitly mentioned this factor in the interviews (e.g., P1 & P6) and highlighted the need for moral considerations in decision making (e.g., P5). Hence, quantitative and qualitative data converge.

Furthermore, the participants recognized that the effects of pollution are global, as most of them rated a high score on statement c. In the interviews, several of them commented that pollution will not simply disappear by relocating polluting sources to other places (e.g., P4, P7, & P8). Therefore, it is reasonable to conclude that the participants expressed consistent views in the interviews and in response to the survey question.

Finally, the participants agreed on statement d, which states that moving polluting industries is not a responsible action and the United States should solve this problem in its own land. Similarly, in their interviews, the participants urged the United States to take responsibility and take action (e.g., P2, P3, & P8). Thus, quantitative and qualitative data match with each other.

The last notable result is participants' ratings on statement e and f. The expert panel considered statement e as a "plausible" one; however, most participants agreed on this statement. Participants' high rating on this statement lowers their ultimate index, because high rating on a "plausible" statement is less viable. Conversely, the participants had varied views regarding statement f, which was assigned as "appropriate" by the expert panel. Some participants totally agreed with it, some moderately agreed with it, and some disagreed with it. Due to the ratings on statement e and f, some participants had a moderate total index on item 40161, although they were able to express proper views on other statements. Overall, participants' survey responses to item 40161 matched with their positions expressed in the interviews.

TABLE 5.6 Participants' Rating on Every Statement of Item 40161 and His/Her Total Index

Item 40161. Heavy industry has greatly polluted North America. Therefore, it is a responsible decision to move heavy industry to underdeveloped countries where pollution is not so widespread.	Expert Panel	Participant's Rating							
	A/P/N	P1	P2	P3	P4	P5	P6	P7	P8
Heavy industry should be moved to underdeveloped countries to save our country and its future generations from pollution.	N	2	1	1	1	2	1	1	2
It is hard to tell. By moving industry we would help poor countries to prosper and we would help reduce our own pollution. But we have no right to pollute someone else's environment.	A	8	8	9	9	7	9	9	9
It doesn't matter where industry is located. The effects of pollution are global.	A	7	9	8	9	6	9	9	9
Heavy industry should NOT be moved to underdeveloped countries.									
Because moving industry is not a responsible way of solving pollution, we should reduce or eliminate pollution here, rather than create more problems elsewhere.	A	8	9	9	9	6	9	9	9
Because those countries have enough problems without the added problem of pollution.	P	8	8	9	5	5	9	9	9
Because pollution should be confined as much as possible. Spreading it around would only create more damages.	A	5	6	9	9	2	9	9	5
	Index	.25	.42	.31	1.00	.60	.33	.33	.17

Note: The numbers represent the agreement between each statement and the participant's opinion, with agreement increasing with the number. The corresponding scaling for the A/P/N positions and the computation of an index are summarized in Vazquez-Alonso and Manassero-Mas (1999, p. 242).

The Role of Science and Technology in Moral Decisions versus Item 40221

Interview transcripts are compared to participants' survey responses to item 40221. Table 5.7 summarizes participants' rating on every statement of item 40221, which examines the relationship between moral decisions and science and technology. When examining the quantitative and qualitative data together, there are several remarkable features to emphasize.

First, according to Table 5.7, Participant 5 and Participant 7 gave low scores on statement a, which indicates that they disagreed with the argument that information provided by science and technology can help people make moral decisions. Although their positions were different with that of

TABLE 5.7 Participants' Rating on Every Statement of Item 40221 and His/Her Total Index

Item 40221. Science and technology can help people make some moral decisions.	Expert Panel A/P/N	Participant's Rating							
		P1	P2	P3	P4	P5	P6	P7	P8
By making you more informed about people and the world around you. This background information can help you cope with the moral aspects of life.	A	6	8	9	6	4	9	1	8
By providing information; but moral decisions must be made by individuals.	A	6	6	4	9	5	9	9	8
Because science includes areas like psychology which study the human mind and emotions.	N	5	7	4	7	5	9	1	8
Science and technology CANNOT help people make a moral decision.									
Because science and technology have nothing to do with moral decisions. Science and technology only discover, explain, and invent things. What people do with the results is not scientists' concern.	N	2	8	1	9	3	1	9	5
Because moral decisions are made solely on the basis of an individual's values and beliefs.	N	2	7	1	9	3	1	1	5
Because if moral decisions are based on scientific information, the decisions often lead to racism, by assuming that one group of people is better than another group.	N	2	8	6	5	2	1	1	5
	Index	.41	−.06	.44	.31	.16	.75	.25	.28

Note: The numbers represent the agreement between each statement and the participant's opinion, with agreement increasing with the number. The corresponding scaling for the A/P/N positions and the computation of an index are summarized in Vazquez-Alonso and Manassero-Mas (1999, p. 242).

the expert panel's, they expressed same concerns in the interviews. For example, although Participant 5 acknowledged that science and technology can provide information, she believes that science does not tell us what is moral and stated, "I don't think science tells us what is moral, but science tells us the facts." Similarly, Participant 7 expressed great concerns regarding the way that technology is currently used and stated that technology "is more produced for consumption and disposal." Both of them believe that science and technology alone cannot help people make moral decisions. Overall, their positions in the interviews and in the survey are consistent.

Second, Participant 4 totally agreed on statements d and e, which argue that moral decisions are made by individuals, and therefore, science and technology do not have a prominent role in moral decisions. His survey response is supported by his interview discussion, when he discussed his grandma who chose to die naturally and refused to use machines and medicines under a very ill condition. Even though his family knows that science can prolong his grandma's life, they decided to respect her decision and let her determine how she wants to live. From his experience, he believes that science sometimes can pose conflicts with traditional beliefs. It is fair to conclude that his survey response converge with his interview discussion.

Finally, it is interesting to note that the participants with science and technology backgrounds (i.e., P2, P4, P5, & P7) had relatively low total index on item 40221 compared to the rest of the participants without adequate science and technology experiences (i.e., P1, P3, P6, & P8). It seems that the participants with science and technology backgrounds have a higher level of concern and a lower belief in the benefits of science. By contrast, the participants without adequate science and technology backgrounds tend to have a strong belief in the promise of science and a low concern of possible dangers. It is possible that people in the field of science and technology contemplate the benefits and dangers of science frequently, and hence, are aware that science and technology may, at times, pose danger to traditional moral belief systems. However, people outside the science field usually have a progressive stereotype about science and technology, and hence, generally react positively to science-related issues.

DISCUSSION

Unidimensionality Assumption of the MRM

Vazquez-Alonso and Manassero-Mas (1999) assert that the MRM guarantees unidimensionality for a set of VOSTS items. They argue that the measured construct is the attitudes' degree of appropriateness with respect to an adequate conception of the science (certified by the history, the epistemology, and the sociology of science and implemented through the categories assigned to positions by the expert panel), and therefore, every item of this construct is consistent. Consequently, they conclude that "the individual consistence on every item of this construct guarantees the unidimensionality for a set of items" (p. 243). However, based upon the findings from the current study regarding the e-waste issue, we argue that the MRM does not provide *sufficient* evidence of unidimensionality.

First, the concepts of internal consistency and unidimensionality are different in nature; they appear to be very similar, but they are certainly not the same. Internal consistency tests the question of whether items share variances or not. That is, if items share variances (no matter with "some" others or "all" items), they are internally consistent (Gardner, 1995). However, unidimensionality explores a different question of whether *all* items share variances. The words "some" and "all" are of great significance in the unidimensionality discussion.

To explain, only if all items share common variances is this set of items unidimensional (Gardner, 1995). Therefore, internal consistency does not provide sufficient evidence of unidimensionality. In other words, the only necessary condition for internal consistency is that items share variances with *some* other items in the scale. By contrast, a unidimensional construct requires that a set of items *all* load on a common factor. In general, if a scale is unidimensional, it follows that it will be internally consistent. The converse, however, is not necessarily true: A scale can be internally consistent and display a high Cronbach's alpha value, but this does not necessarily mean that it is unidimensional (Gardner, 1995). Therefore, it is debatable to claim that "the individual consistence on every item of this construct guarantees the unidimensionality for a set of items" (Vazquez Alonso & Manassero-Mas, 1999, p. 243).

Second, although expert judges' views are generally certified by the history, the epistemology, and the sociology of science, their adequacy of knowledge can vary across a number of VOSTS items. Rubba and colleagues (1996) examined variance among five expert judges' categorizations across 16 VOSTS items. They found a statistically significant difference between one judge's assignment and the other four judges'. In general, we do not agree with Vazquez-Alonso and Manassero-Mas's (1999) argument that consistence on every item of the measured construct is certified by the categories assigned to positions. We think the meaning of every position is defined locally, within the context and topic of each individual item. A set of items can test different dimensions of STS conceptions, and consequently, the MRM does not account for this multidimensional or contextual nature for a set of items. Thus, we argue that multidimensionality is found among the test items that were selected from the VOSTS item pool and used in this study.

Overall, it is important to examine the MRM's internal reliability and dimensionality while applying it in an empirical setting. Although the VOSTS item pool per se is qualitative in nature, the MRM lends to the instrument a Likert scale. Now, the instrument has a 9-point semantic differential scale with parametric scores that are addable. The MRM makes the VOSTS instrument a psychometric instrument that measures cognitive, affective, and psychomotor attributes, and subsequently, yields quantitative constructs. The MRM meets the continuity and equal intervals of measures assumptions (Kerlinger, 1973) and needs to assess the reliability of a set of selected VOSTS items.

Limitations of the MRM

The MRM turns the VOSTS instrument into a new version of a Likert questionnaire. It enables hypothesis testing and inferential statistics. However, it has the following limitations.

First, the selection of VOSTS items matters. In the study described herein, the selection was based on the degree of relevance to the e-waste issue. Due to the exploratory nature of the study, we expected to broadly investigate the influence of STS conceptions on decision making. Therefore, items from different sections of the VOSTS instrument were chosen. The number of items was restricted to seven mainly due to the instrument administration time. Reading positions and then assessing them on the attached rating scale is time consuming, even in the administration of an online instrument that was used in this study.

The construct about STS conceptions was, therefore, defined by these seven items. It can be argued that by selecting one different item, we would have defined the construct of STS conceptions differently and probably would have gotten different results. For example, rather than selecting seven items from seven topics as implemented in this study, selecting more than one item from the same topic could increase inter-item correlation and so, probably, would impact research findings.

Second, the A/P/N categorization of positions forms another limitation. Due to variance in views among the panel members, obtaining a majority decision turned out to be difficult on a number of positions. In this study, several expert judges remarked their uncertainty about several positions. Rubba and colleagues (1996) speculate that this is because STS interactions are not a topic frequently thought about, even among scientists. The A/P/N categorization limits the generalizability of research findings to other socio-scientific topics or different research contexts. To solve this problem, the researchers (Rubba et al., 1996) suggest the use of at least nine expert judges to achieve agreement among at least seven of the nine to be used as the criterion. As reported, we intervened in the cases when disagreement happened. Out of a total of 36 VOSTS positions from the seven VOSTS items of this study, we determined the final categorizations of four positions. It can be debated that having assigned positions into different categories would lead to different research findings; however, the qualitative data supports those positions as being appropriate for the participants.

Finally, employing the MRM does not guarantee objectivity per se. In fact, every study is generally subject to researchers' subjectivity. For example, this study investigated how participants as customers recognize their voices in advancing "green" electronic products and is subject to our opinions on this issue. As the researchers, we view the following VOSTS position as "appropriate": "Technological developments can be controlled by citizens, because technology serves the needs of consumers. Technological developments will occur in areas of high demand and where profits can be made in the marketplace."

We believe consumers' needs will motivate companies to adopt higher standards in their operations. Consequently, the dynamics of buyer–seller interaction will lead to further advancement of technology across the industry. More specifically, consumers' ecological consciousness and corresponding willingness for eco-friendly green purchases is likely to motivate high-tech companies to adopt the concept of green marketing in their operations. For instance, Apple Computer Company responded to the demand of 50,000 Mac fans who wrote to Steve Jobs in May, 2007 and urged Apple to go green. In January 2008, Apple launched the MacBook Air laptop that is mercury and arsenic free. According to Greenpeace's assessment, MacBook Air exceeds European Standards and raises the bar for the rest of the industry. By contrast, the panel's consensus view was against ours: three of the five expert judges assigned the statement to "plausible." Given the purpose of this investigation, we finally categorized the position as "appropriate."

Implications for Future Research

The limitations of the MRM have the following implications for future research. First, since the selection of VOSTS items matters, researchers need to take into consideration the goals of the assessment. It is crucial that item selection be carried out with great care and forethought, since researchers are defining the construct to be measured by selecting relevant items. Indeed, instrument validity depends on careful instrument construction, whether it is in a quantitative or qualitative study (Rubba et al., 1996).

Second, researchers who employ the MRM procedure need to judge the appropriateness of the panel's consensus views against the goals of the study under investigation (Rubba, et al., 1996). If the panel's consensus views do not match with the design of the study, researchers should determine the ultimate assignation of positions (Rubba, et al., 1996). It is important to emphasize that the MRM weights "appropriate" positions more than the remaining positions. As the number of "appropriate" positions is usually a minimum in an item, the procedure of the MRM weights one "appropriate" position the same as the "plausible" or "naive" categories where usually three, four, or more positions are added up. Consequently, the MRM over-scores the contribution of "appropriate" positions; that is, answers on "appropriate" positions get a relative greater weight than the remaining positions. In general, researchers need to use extra discretion when intervening in these cases.

Finally, since the application of the MRM does not ensure objectivity, researchers are suggested to report research findings following not only the quantitative criteria of internal consistency and dimensionality, but also the qualitative criteria of credibility, transferability, dependability, and confirmability. On one hand, the quantitative criteria are required, because the MRM lends to the VOSTS instrument a Likert scale that needs to verify the reliability of selected VOSTS items. On the other hand, the application of the MRM involves researchers' personal biases that call for following strategies to counter. More specifically, credibility examines how rigorous the methods for conducting a study are. Strategies to enhance credibility include making biases explicit, assessing rival conclusions, testing negative cases, peer debriefing, and member checking (Guba & Lincoln, 1989; Patton, 2002). Transferability involves providing an extensive and careful description of the time, the place, the context, and the culture in which hypotheses become relevant so that readers are able to apply findings from this study to their own interests and situations (Guba & Lincoln, 1989). The idea of dependability emphasizes the need for the researcher to account for the ever-changing context within which research occurs (Guba & Lincoln, 1989). The criterion for establishing confirmability is intended to allow readers to understand and evaluate how the data is coded and categorized, why data is placed into these codes and categories, and how these are clustered to answer the research questions (Guba & Lincoln, 1989). In summary, to ensure the trustworthiness of research findings, the MRM is not suggested to be used alone as a substitute for detailed descriptive analysis of STS conceptions (Rubba et al., 1996).

CONCLUSION

The application of a mixed-methods approach enables us to quantitatively analyze VOSTS data and detect statistically significant relationships between STS understanding and socio-scientific decision making, which is a clear improvement in research results compared to other published attempts (Bell & Lederman, 2003; Sadler, Chambers, & Zeidler, 2004; Zeidler et al., 2002). Internal consistency test shows that the selected set of seven VOSTS items has a low Cronbach's alpha of .344 and, hence, lacked internal consistency and was multidimensional. That is, this set tests different dimensions of STS and does not load on a common factor. Consequently, the items were analyzed separately, rather than being summated to yield a total score. A close examination of Vazquez-Alonso and Manassero-Mas's (1999) rationale in light of the empirical findings of this study suggests that internal consistency and unidimensionality are separate entities. Thus, we suggest paying more attention to examining internal reliability and dimensionality while applying the MRM in an empirical setting.

We used qualitative data to triangulate quantitative results. Patterns emerged from interviews matched with respondents' responses to survey questions. For example, regarding the issue of moving polluting industries to underdeveloped countries, the following patterns emerged from the interviews: economic considerations, concerns about global effects of pollution, call for responsibility, and empathy. In accordance, the participants gave full agreements on corresponding survey statements that discuss these perspectives. The convergence of quantitative and qualitative results ensures the trustworthiness of findings and adds strength to research quality. Overall, measuring STS attitude is indeed a complicated and challenging endeavor, yet it is vital to achieve a practical degree of the goal of scientific literacy (AAAS, 1990, 1993; NRC, 1996).

REFERENCES

Aikenhead, G. (1979). Using qualitative data in formative evaluation. *Alberta Journal of Educational Research, 25*(2), 117–129.

Aikenhead, G. S. (1987). High-school graduates' beliefs about science-technology-society. III. Characteristics and limitations of scientific knowledge. *Science Education, 71,* 459–487.

Aikenhead, G. S. (1994). Science curricula and preparation for social responsibility. In J. Solomon & G. S. Aikenhead (Eds.), *STS education: International perspectives on reform* (pp. 47–59). New York: Teachers College Press.

Aikenhead, G. S., & Ryan, A. G. (1992). The development of a new instrument: "Views on Science-Technology-Society" (VOSTS). *Science Education, 76*(5), 477–491.

Aikenhead, G. S., Ryan, A. G., & Fleming, R. G. (1989). *Views on Science-Technology-Society.* Saskatoon, Saskatchewan: University of Saskatchewan, Department of Curriculum Studies.

American Association for the Advancement of Science (AAAS). (1990). *Project 2061: Science for all Americans.* Washington, DC: American Association for the Advancement of Science.

American Association for the Advancement of Science (AAAS). (1993). *Benchmarks for science literacy: A Project 2061 report.* Washington, DC: Oxford University Press.

Bady, R. A. (1979). Students' understanding of the logic of hypothesis testing. *Journal of Research in Science Teaching, 16,* 61–65.

Bell, R., & Lederman, N. (2003). Understandings of the nature of science and decision making on science and technology based issues. *Science Education, 87,* 352–377.

Fleming, R. G. (1988). Undergraduate science students' views on the relationship between science, technology, and society. *International Journal of Science Education, 10*(4), 449–463.

Gardner, P. L. (1995). Measuring attitudes to science: Unidimensionality and internal consistency revisited. *Research in Science Education, 25*(3), 283–289.

Guba, E., & Lincoln, Y. (1989). Judging the quality of fourth generation evaluation. In *Fourth generation evaluation* (pp. 228–251). Newbury Park, CA: Sage.

Kerlinger, F. N. (1973). *Foundations of behavioral research.* New York: Holt, Rinehart and Winston.

Layton, D. (1994). STS in the school curriculum: A movement overtaken by history? In J. Solomon & G. S. Aikenhead (Eds.), *STS education: International perspectives on reform* (pp. 32–46). New York: Teachers College Press.

Lederman, N. G. (1992). Students' and teachers' conceptions of the nature of science: A review of the research. *Journal of Research in Science Teaching, 29,* 331–359.

Miller, P. E. (1963). A comparison of the abilities of secondary teachers and students of biology to understand science. *Iowa Academy of Science, 70,* 510–513.

National Research Council (NRC). (1996). *National science education standards.* Washington, DC: National Academic Press.

Patton, M. Q. (1990). *Qualitative research and evaluation methods* (2nd ed.). Thousand Oaks, CA: Sage Publications.

Rubba, P. A., Bradford, C. S., & Harkness, W. J. (1996). A new scoring procedure for the Views On Science-Technology-Society instrument. *International Journal of Science Education, 18*(4), 387–400.

Rubba, P., Horner, J., & Smith, J. M. (1981). A study of two misconceptions about the nature of science among junior high school students. *School Science and Mathematics, 81,* 221–226.

Saphores, J. M., Nixon, H., Ogunseitan, O. A., & Shapiro, A. A. (2006). Household willingness to recycle electronic waste: An application to California. *Environment and Behavior, 38*(2), 183–208.

Strauss, A. L., & Corbin, J. (1998). *Basics of qualitative research: Techniques and procedures for developing grounded theory* (2nd ed.). Thousand Oaks, CA: Sage Publications, Inc.

Vazquez-Alonso, A., & Manassero-Mas, M. (1999). Response and scoring models for the "Views on Science-Technology-Society" instrument. *International Journal of Science Education, 21*(3), 231–247.

Yang, F., & Anderson, O. R. (2003). Senior high school students' preference and reasoning modes about nuclear energy use. *International Journal of Science Education, 25*(2), 221–244.

Zeidler, D. L., Walker, K. A., Ackett, W. A., & Simmons, M. L. (2002). Tangled up in views: Beliefs in the nature of science and responses to socioscientific dilemmas. *Science Education, 86*(3), 343–367.

Ziman, J. (1994). The rationale of STS education is in the approach. In J. Solomon & G. S. Aikenhead (Eds.), *STS education: International perspectives on reform* (pp. 21–31). New York: Teachers College Press.

Zoller, U., Ebenezer, J., Morely, K., Paras, S., Sandberg, V., West, C., et al. (1990). Goal attainment in science-technology-society (STS) education and reality: The case of British Columbia. *Science Education, 74*(1), 19–36.

APPENDIX A
The E-Waste Survey

Introduction

In this study, we are trying to investigate our knowledge about electronic waste and explore your beliefs about the nature of science and technology. We are interested in your decision making about a number of issues that relate to the electronic waste issue.

The following survey requires 15–20 minutes to complete. Please respond to all of the questions. Thank you for your participation!

Section I: Demographic information

1. Your name:
2. Email:
3. Gender:
4. Department and program:
5. Are you an international student?

Section II: Views on Science-Technology-Society

In every statement, please circle the number that best represents the agreement between your opinion on the issue and the position expressed in the statement on a scale of 1 to 9. Agreement increases with the numbers.

Q1. *Technological developments can be controlled by citizens.*	**Your agreement:** **Agreement increasing** **with the numbers**
a) Yes, because from the citizen population comes each generation of the scientists and technologists who will develop the technology. Thus citizens slowly control the advances in technology through time.	1 2 3 4 5 6 7 8 9
b) Yes, because technological advances are sponsored by the government. By electing the government, citizens can control what is sponsored.	1 2 3 4 5 6 7 8 9
c) Yes, because technology serves the needs of consumers. Technological developments will occur in areas of high demand and where profits can be made in the market place.	1 2 3 4 5 6 7 8 9
d) Yes, but only when it comes to putting new developments into use. Citizens cannot control the original development itself.	1 2 3 4 5 6 7 8 9
e) Yes, but only when citizens get together and speak out either for or against a new development. Organized people can change just about anything.	1 2 3 4 5 6 7 8 9
No, citizens are NOT involved in controlling technological developments:	
f) Because technology advances so rapidly that the average citizen is left ignorant of the development.	1 2 3 4 5 6 7 8 9
g) Because citizens are prevented from doing so by those with the power to develop the technology.	1 2 3 4 5 6 7 8 9

Q2. *When a new technology is developed (for example, a new computer), it may or may not be put into practice. The decision to use a new technology depends on whether the advantages to society outweigh the disadvantages to society.*

Your agreement: Agreement increasing with the numbers

a) The decision to use a new technology depends mainly on the benefits to society because if there are too many disadvantages, society won't accept it and may discourage further development.

1 2 3 4 5 6 7 8 9

b) The decision depends on more than just the technology's advantages and disadvantages. It depends on how well it works, its cost, and its efficiency.

1 2 3 4 5 6 7 8 9

c) It depends on your point of view. What is an advantage to some people may be a disadvantage to others.

1 2 3 4 5 6 7 8 9

d) Many new technologies have been put into practice to make money and gain power, even though their disadvantages were greater than their advantages.

1 2 3 4 5 6 7 8 9

Q3. *Heavy industry has greatly polluted North America. Therefore, it is a responsible decision to move heavy industry to underdeveloped countries where pollution is not so widespread.*

Your agreement: Agreement increasing with the numbers

a) Heavy industry should be moved to underdeveloped countries to save our country and it future generations from pollution.

1 2 3 4 5 6 7 8 9

b) It is hard to tell. By moving industry we would help poor countries to prosper and we would help reduce our own pollution. But we have no right to pollute someone else's environment.

1 2 3 4 5 6 7 8 9

c) It doesn't matter where industry is located. The effects of pollution are global.

1 2 3 4 5 6 7 8 9

Heavy industry should NOT be moved to underdeveloped countries.

d) Because moving industry is not a responsible way of solving pollution, we should reduce or eliminate pollution here, rather than create more problems elsewhere.

1 2 3 4 5 6 7 8 9

e) Because those countries have enough problems without the added problem of pollution.

1 2 3 4 5 6 7 8 9

f) Because pollution should be confined as much as possible. Spreading it around would only create more damages.

1 2 3 4 5 6 7 8 9

Q4. *Scientists and engineers should be the ones to decide on North American air pollution standards (for example, industrial emissions of sulfur dioxide, pollution control gadgets for your car or truck, sour gas emissions from oil well, etc.) because scientists and engineers are the people who know the facts best.*

Your agreement: Agreement increasing with the numbers

a) Because they have the training and facts which give them a better understanding of the issue.
1 2 3 4 5 6 7 8 9

b) Because they have the knowledge and can make better decisions than government bureaucrats or private companies, both of whom have vested interested.
1 2 3 4 5 6 7 8 9

c) Because they have the training and facts which give them a better understanding. But the public should be involved—either informed or consulted.
1 2 3 4 5 6 7 8 9

d) The decision should be made equally; viewpoints of scientists and engineers, other specialists, and the informed public should all be considered in decisions which affect our society.
1 2 3 4 5 6 7 8 9

e) The government should decide because the issue is basically a political one. BUT scientists and engineers should give advice.
1 2 3 4 5 6 7 8 9

f) The public should decide because the decision affects everyone. BUT scientists and engineers should give advice.
1 2 3 4 5 6 7 8 9

g) The public should decide because the public serves as a check on the scientists and engineers. Scientists and engineers have idealistic and narrow views on the issue and thus pay little attention to consequences.
1 2 3 4 5 6 7 8 9

Q5. *Science and technology can help people make some moral decisions (that is, one group of people deciding how to act towards another group of people).*

Your agreement: Agreement increasing with the numbers

a) By making you more informed about people and the world around you. This background information can help you cope with the moral aspects of life.
1 2 3 4 5 6 7 8 9

b) By providing information; but moral decisions must be made by individuals.
1 2 3 4 5 6 7 8 9

c) Because science includes areas like psychology which study the human mind and emotions.
1 2 3 4 5 6 7 8 9

Science and technology CANNOT help people make a moral decision.

d) Because science and technology have nothing to do with moral decisions. Science and technology only discover, explain, and invent things. What people do with the results is not scientists' concern.
1 2 3 4 5 6 7 8 9

e) Because moral decisions are made solely on the basis of an individual's values and beliefs.
1 2 3 4 5 6 7 8 9

f) Because if moral decisions are based on scientific information, the decisions often lead to racism, by assuming that one group of people is better than another group.
1 2 3 4 5 6 7 8 9

Q6. *We always have to make tradeoffs (compromises) between the positive and negative effects of science and technology.*

Your agreement: Agreement increasing with the numbers

a) Because every new development has at least one negative result. If we didn't put up with the negative results, we would not progress to enjoy the benefits.
1 2 3 4 5 6 7 8 9

b) Because scientists cannot predict the long-term effects of new developments, in spite of careful planning and testing. We have to take the chance.
1 2 3 4 5 6 7 8 9

c) Because things that benefit some people will be negative for someone else. This depends on a person's viewpoint.
1 2 3 4 5 6 7 8 9

d) Because you cannot get positive results without first trying a new idea and then working out its negative effects.
1 2 3 4 5 6 7 8 9

e) But the trade-offs make no sense. (For example: Why should inventing labor-saving devices cause more unemployment? Or Why should defending a country with nuclear weapons threaten life on earth?)
1 2 3 4 5 6 7 8 9

There are NOT always trade-offs between benefits and negative effects.

f) Because some new developments benefit us without producing negative effects.
1 2 3 4 5 6 7 8 9

g) Because negative effects can be minimized through careful planning and testing.
1 2 3 4 5 6 7 8 9

h) Because negative effects can be eliminated through careful planning and testing. Otherwise, a new development is not used.
1 2 3 4 5 6 7 8 9

Q7. *We have to be concerned about pollution problems that are unsolvable today. Science and technology cannot necessarily fix these problems in the future.*

Your agreement: Agreement increasing with the numbers

a) Because science and technology are the reason that we have pollution problems in the first place. More science and technology will bring more pollution problems.
1 2 3 4 5 6 7 8 9

b) Because pollution problems are so bad today they are already beyond the ability for science and technology to fix them.
1 2 3 4 5 6 7 8 9

c) Because pollution problems are BECOMING so bad that they may soon be beyond the ability of science and technology to fix them.
1 2 3 4 5 6 7 8 9

d) No one can predict what science and technology will be able to fix in the future.
1 2 3 4 5 6 7 8 9

e) Science and technology alone cannot fix pollution problems. It is everyone's responsibility. The public must insist that fixing these problems is a top priority.
1 2 3 4 5 6 7 8 9

f) Science and technology can fix such problems because the success at solving problems in the past means science and technology will be successful in the future at fixing pollution problems.
1 2 3 4 5 6 7 8 9

Note: The items have their corresponding numbers in the VOSTS instrument as the following.

Number in the Survey	Number in the VOSTS instrument
Q1	80211
Q2	80131
Q3	40161
Q4	40212
Q5	40221
Q6	40311
Q7	40451

Section III: Please rate your willingness on a scale of 1 to 5.

1. Would you be willing to pay extra tax for environmental protection?
   (1 = very unwilling; 5 = very willing)
       1   2   3   4   5

2. Would you be willing to pay more for environmentally friendly products?
   (1 = very unwilling; 5 = very willing)
       1   2   3   4   5

3. Would you be willing to pay extra tax to provide research funding for greener design of electronic products?
   (1 = very unwilling; 5 = very willing)
       1   2   3   4   5

4. Would you be willing to pay extra for a green cell phone that is mercury-free?
   (1 = very unwilling; 5 = very willing)
       1   2   3   4   5

5. Would you be willing to pay extra for electronic waste recycling?
   (1 = very unwilling; 5 = very willing)
       1   2   3   4   5

6. Would you be willing to sign petition that demands high-tech companies abandon hazardous substances in their products?
   (1 = very unwilling; 5 = very willing)
       1   2   3   4   5

APPENDIX B
Interview Protocols

Interview Questions
**Introduce myself and this study

1. Tell me about yourself: your program, the stage of your study, your research interest, why you agreed to participate in this interview.
2. Do you believe that science and technology can help people make moral decisions? If so, in what specific ways?
3. Can you provide some examples in your life that science/technology helped you make moral decisions or failed to help make moral decisions?
4. I would like for you to watch this short 4.5 minute video and then we will discuss it— E-waste: Dump on the poor (http://www.youtube.com/watch?v=EXzsqTFwV3Q&feature=related)
5. What are your first impressions from watching the video? Which part of the video opened your views, if at all, to the e-waste problem locally, nationally and globally?
6. The US was moving heavy industrial wastes to underdeveloped countries? Do you think this is an acceptable practice?
7. How do you dispose of your electronic wastes (e.g. computers, cell phones, etc.)? As consumers of electronics, what can we do to help people become more aware of the issue of e-wastes? What can we do to protect poor countries, or the environment?
8. As students and educators, what can we do to learn and to educate others about the environment and/or electronic waste issues?
9. What have you learned, if anything from participating in this survey you took a few months ago?
10. What information do you think is important that helps you understand the e-waste issue better? What information do you think is missing that will help your understanding?

**I really appreciate your time and interest in my study. Can I send you my tentative data interpretations for member checking? Get email address.

APPENDIX C
Expert Judges' Categorization of the VOSTS Items

Please categorize **each statement** in a VOSTS item to one of the following three categories with respect to the current knowledge of history, epistemology, and sociology of science:

A/Appropriate: The statement expresses an appropriate view.
P/Plausible: While not completely appropriate, the statement expresses some plausible points.
N/Naïve: The statement expresses a view that is inappropriate or not plausible.

Note: The shaded letters are the experts' categorization.

Q1. *Technological developments can be controlled by citizens.*	Your categorization
a) Yes, because from the citizen population comes each generation of the scientists and technologists who will develop the technology. Thus citizens slowly control the advances in technology through time.	A P N
b) Yes, because technological advances are sponsored by the government. By electing the government, citizens can control what is sponsored.	A P N
c) Yes, because technology serves the needs of consumers. Technological developments will occur in areas of high demand and where profits can be made in the market place.	A P N
d) Yes, but only when it comes to putting new developments into use. Citizens cannot control the original development itself.	A P N
e) Yes, but only when citizens get together and speak out either for or against a new development. Organized people can change just about anything.	A P N
No, citizens are NOT involved in controlling technological developments:	
f) Because technology advances so rapidly that the average citizen is left ignorant of the development.	A P N
g) Because citizens are prevented from doing so by those with the power to develop the technology.	A P N

Q2. *When a new technology is developed (for example, a new computer), it may or may not be put into practice. The decision to use a new technology depends on whether the advantages to society outweigh the disadvantages to society.*

Your categorization

a)	The decision to use a new technology depends mainly on the benefits to society because if there are too many disadvantages, society won't accept it and may discourage further development.	A	**P**	N
b)	The decision depends on more than just the technology's advantages and disadvantages. It depends on how well it works, its cost, and its efficiency.	**A**	P	N
c)	It depends on your point of view. What is an advantage to some people may be a disadvantage to others.	A	**P**	N
d)	Many new technologies have been put into practice to make money and gain power, even though their disadvantages were greater than their advantages.	A	**P**	N

Q3. *Heavy industry has greatly polluted North America. Therefore, it is a responsible decision to move heavy industry to underdeveloped countries where pollution is not so widespread.*

Your categorization

a)	Heavy industry should be moved to underdeveloped countries to save our country and its future generations from pollution.	A	P	**N**
b)	It is hard to tell. By moving industry we would help poor countries to prosper and we would help reduce our own pollution. But we have no right to pollute someone else's environment.	**A**	P	N
c)	It doesn't matter where industry is located. The effects of pollution are global.	**A**	P	N

Heavy industry should NOT be moved to underdeveloped countries.

d)	Because moving industry is not a responsible way of solving pollution, we should reduce or eliminate pollution here, rather than create more problems elsewhere.	**A**	P	N
e)	Because those countries have enough problems without the added problem of pollution.	A	**P**	N
f)	Because pollution should be confined as much as possible. Spreading it around would only create more damages.	**A**	P	N

Q4. *Scientists and engineers should be the ones to decide on North American air pollution standards (for example, industrial emissions of sulfur dioxide, pollution control gadgets for your car or truck, sour gas emissions from oil well, etc.) because scientists and engineers are the people who know the facts best.*

Your categorization

a) Because they have the training and facts which give them a better understanding of the issue.   A   P   N

b) Because they have the knowledge and can make better decisions than government bureaucrats or private companies, both of whom have vested interested.   A   P   N

c) Because they have the training and facts which give them a better understanding. But the public should be involved—either informed or consulted.   A   P   N

d) The decision should be made equally; viewpoints of scientists and engineers, other specialists, and the informed public should all be considered in decisions which affect our society.   A   P   N

e) The government should decide because the issue is basically a political one. BUT scientists and engineers should give advice.   A   P   N

f) The public should decide because the decision affects everyone. BUT scientists and engineers should give advice.   A   P   N

g) The public should decide because the public serves as a check on the scientists and engineers. Scientists and engineers have idealistic and narrow views on the issue and thus pay little attention to consequences.   A   P   N

Q5. *Science and technology can help people make some moral decisions (that is, one group of people deciding how to act towards another group of people.*

Your categorization

a) By making you more informed about people and the world around you. This background information can help you cope with the moral aspects of life.   A   P   N

b) By providing information; but moral decisions must be made by individuals.   A   P   N

c) Because science includes areas like psychology which study the human mind and emotions.   A   P   N

Science and technology CANNOT help people make a moral decision.

a) Because science and technology have nothing to do with moral decisions. Science and technology only discover, explain, and invent things. What people do with the results is not scientists' concern.   A   P   N

b) Because moral decisions are made solely on the basis of an individual's values and beliefs.   A   P   N

c) Because if moral decisions are based on scientific information, the decisions often lead to racism, by assuming that one group of people is better than another group.   A   P   N

Q6. *We always have to make tradeoffs (compromises) between the positive and negative effects of science and technology.*

Your categorization

a)	Because every new development has at least one negative result. If we didn't put up with the negative results, we would not progress to enjoy the benefits.	A P N
b)	Because scientists cannot predict the long-term effects of new developments, in spite of careful planning and testing. We have to take the chance.	A P N
c)	Because things that benefit some people will be negative for someone else. This depends on a person's viewpoint.	A P N
d)	Because you cannot get positive results without first trying a new idea and then working out its negative effects.	A P N
e)	But the trade-offs make no sense. (For example: Why should inventing labor-saving devices cause more unemployment? Or Why should defending a country with nuclear weapons threaten life on earth?)	A P N

There are NOT always tradeoffs between benefits and negative effects.

f)	Because some new developments benefit us without producing negative effects.	A P N
g)	Because negative effects can be minimized through careful planning and testing.	A P N
h)	Because negative effects can be eliminated through careful planning and testing. Otherwise, a new development is not used.	A P N

Q7. *We have to be concerned about pollution problems that are unsolvable today. Science and technology cannot necessarily fix these problems in the future.*

Your categorization

a)	Because science and technology are the reason that we have pollution problems in the first place. More science and technology will bring more pollution problems.	A P N
b)	Because pollution problems are so bad today they are already beyond the ability for science and technology to fix them.	A P N
c)	Because pollution problems are BECOMING so bad that they may soon be beyond the ability of science and technology to fix them.	A P N
d)	No one can predict what science and technology will be able to fix in the future.	A P N
e)	Science and technology alone cannot fix pollution problems. It is everyone's responsibility. The public must insist that fixing these problems is a top priority.	A P N
f)	Science and technology can fix such problems because the success at solving problems in the past means science and technology will be successful in the future at fixing pollution problems.	A P N

CHAPTER 6

TAILORING INFORMATION TO CHANGE ATTITUDES

A Meta-Structural Approach

Ya Hui Michelle See and Bernice L. Z. Khoo
National University of Singapore

ABSTRACT

The assessment of attitudes toward science education is important in order to design messages and teaching strategies that will promote positivity toward studying science. A useful approach is to consider the bases of attitudes from both a structural perspective and a meta-attitudinal perspective. For instance, affective or cognitive structural bases of attitudes represent associations among emotions, beliefs, and overall evaluations in one's memory, whereas affective or cognitive meta-bases of attitudes refer to subjective judgments about one's evaluations being driven primarily by emotions or beliefs. Results from a pilot study suggest differences in affective and cognitive structural bases and meta-bases for attitudes toward physics, chemistry, and biology. The meta-structural distinction can also be extended to include other types of attitudinal bases such as value-expressive and social-adjustive functions, one's group members, one's self-schema, and so forth. The meta-structural distinction has consequences for persuasion, as recent research suggests that information that is

Attitude Research in Science Education, pages 177–198

tailored to match an individual's meta- and structural attitudinal features has consequences for the motivation and efficiency for processing the matched information, respectively. Implications for designing messages to increase interest in science education will be discussed.

In order to gauge a student's interest in studying science or pursuing a scientific career, it is important for an educator to assess the student's attitude toward science education. An attitude refers to an overall evaluation toward an object, where the object could be an abstract entity (e.g., scientific knowledge), a product (e.g., calculators), or a person (e.g., my physics teacher).

Various frameworks have been used to define and conceptualize attitudes (see Eagly & Chaiken, 1993; Gawronski, 2007). One relevant framework is the tripartite model, which examines attitudes as driven by one or a combination of three classes of information—affective, cognitive, and behavioral (e.g., Katz & Stotland, 1959; Rosenberg & Hovland, 1960; Zanna & Rempel, 1988). Empirical evidence for the tripartite model has been provided by past research, in which physiological measures of the sympathetic nervous system and self-report mood measures assessed affective bases, measures of overt action and behavioral intention assessed behavioral bases, and self-reports of perceptions and beliefs toward the attitude object assessed cognitive bases (Breckler, 1984). Furthermore, reliable and valid scales that assess the affective and cognitive bases of attitudes across various objects have been developed (Crites, Fabrigar, & Petty, 1994). The objects examined in this research formed a diverse range—literature classes, math classes, capital punishment, birth control, church, and snakes. In other research, affective and cognitive bases of attitudes were measured following expectancy-value models of attitudes (e.g., Fishbein & Ajzen, 1975), in which participants generated their emotions and beliefs toward the target object and then reported the perceived likelihood and valence for each individual emotion and belief (e.g., Eagly, Mladinic, & Otto, 1994; Esses, Haddock, & Zanna, 1993).

The notion that an attitude consists of various bases is also present in existing research on attitudes toward science, where studies that aim to measure attitudes have incorporated different attitudinal bases such as anxiety toward science and perceived value of science (see Osborne, Simon, & Collins, 2003, and Simpson & Oliver, 1990 for reviews). One potential challenge is that such components are viewed as interchangeable ways of measuring the overall attitude but they are more appropriately considered as attitudinal bases that are separate from one another *and* from the global evaluation itself. In other words, the affect-cognition distinction, which has been established to be useful for studying attitudes toward objects such as math classes and snakes, might also be useful for researching attitudes toward science education. For example, one's attitude toward studying physics would be measured by overall positivity toward physics, and one's affective and cognitive bases for the global evaluation would be assessed by excitement about studying physics and perceived value of studying physics, respectively. In addition, the closer in valence emotions are to the overall attitude, the more affective the attitudinal basis. In summary, an attitude can be considered as a mental structure that consists of the overall evaluation itself as well as the various components associated with the evaluation.

ATTITUDE STRUCTURE AND PERSUASION

Although it is possible for an attitude to be equally based on affect and cognition, it is more likely that an attitude is dominated by one (whether it is affect or cognition) rather than the other. The dominant basis of an attitude can be considered at different levels: (1) the attitude object, (2) the individual difference, and (3) the individual difference regarding a specific attitude object. At the attitude object level, we consider whether an object elicits mainly affective or cognitive attitudinal bases across individuals. For example, do people in general base their favorability toward studying physics more on their feelings or on their beliefs? At the individual difference level, we consider whether a person typically holds affectively or cognitively based attitudes across a variety of objects. For instance, does person X rely more on emotions or beliefs in forming preferences, regardless of what the object is? Finally, the dominant basis of an attitude can also be considered with respect to a specific individual and a particular object. For example, does person X rely more on emotions or beliefs in attitudes toward studying physics? Perhaps because of the advantage of generalizability, past research has focused on examining attitudinal bases at the level of the object or at the level of individual differences.

Why is it important to consider the dominant basis of an attitude in persuasion? Much research has shown that matching a persuasive message to one's dominant attitudinal basis increases the effectiveness of the message as long as the message contains strong arguments (see Petty, Wheeler, & Bizer, 2000 for a review). In the affect-cognition domain, affective information was more persuasive than cognitive information when the attitude was dominated by emotions than beliefs, whereas the opposite was true when the attitude was based more on beliefs than on emotions. This effect was observed at the attitude object level, for instance, where participants were induced to hold affective or cognitive evaluations toward a fictitious animal (Fabrigar & Petty, 1999). Similar matching effects were also observed at the individual difference level, for instance, where participants' chronic affective and cognitive bases predicted their susceptibility to emotions-focused versus beliefs-focused appeals regarding a novel sports drink (Huskinson & Haddock, 2004).

In the research mentioned above, emphasis was placed on objective indicators of affective and cognitive attitudinal bases. In other words, the assessment reflected the relationship between the feelings or the beliefs about an object and the overall evaluation toward the object in one's mental representations. Participants knew that they were reporting their emotions and their beliefs, in addition to their global attitudes, toward an attitude object (or multiple attitude objects). However, they did not necessarily know that they were also revealing the respective extents to which their emotions and their beliefs were associated with their evaluations. Put differently, participants were not asked explicitly for their perception of the degree to which their attitudes were based on affect and cognition. As with the objective indicators of affective and cognitive attitudinal bases, subjective perceptions of affective and cognitive attitudinal bases could also differ at the level of the attitude object, the individual, or the individual regarding a specific attitude object. In fact, recent research has established that at the individual difference level, such subjective perceptions (i.e., meta-bases) are not only independent of their objective counterparts (i.e., structural bases), but also have predictive utility for attitudes and related processes (See, Petty, & Fabrigar, 2008).

THE META-STRUCTURAL DISTINCTION IN ATTITUDES

In general, the main approach in attitudes research has been to use subjective and objective measures of various attitudinal properties interchangeably. However, some theorists have suggested that the distinction between the two types of measures may be conceptual and not just methodological (e.g., Bassili, 1996; Krosnick & Petty, 1995). For instance, research examining the role of knowledge in attitude-behavior consistency has used subjective and objective assessments of knowledge while assuming that the effects produced by the two types of knowledge measures are driven by the same underlying mechanism (Davidson, Yantis, Norwood, & Montano, 1985). However, it is possible that subjective knowledge moderates attitude-behavior consistency by influencing the motivation to retrieve attitudes-related information, whereas objective knowledge does the same but by influencing the ability to retrieve attitudes-relevant information.

In research on attitudinal knowledge, subjective assessments are derived from participants' self-reports of how knowledgeable they estimate themselves to be regarding an object whereas objective assessments are derived from participants' actual demonstration of knowledge (e.g., counting the number of pieces of information one generates about the object). Similarly, research on attitudinal ambivalence has also used subjective measures such as participants' self-reports of how conflicted they feel about an object, and objective measures such as the number of positive and negative reactions toward the same object, with the assumption that the two measures are assessing one construct (e.g., Priester & Petty, 1996; Thompson, Zanna, & Griffin, 1995). However, it is worth noting that responses on subjective measures are usually only modestly correlated with objective measures (e.g., Krosnick, Boninger, Chuang, Berent, & Carnot, 1993; Priester & Petty, 1996; Thompson et al. 1995; Wood, Rhodes, & Biek, 1995)

More direct evidence for the conceptual distinction between subjective and objective measures was investigated in three recent studies that examined affective and cognitive bases of attitudes at the individual difference level (See, Petty, & Fabrigar, 2008). Subjective perceptions of bases, or meta-bases, were assessed by participants' self-reports of the extent to which their attitudes toward various objects are driven by emotions versus beliefs. Structural bases were measured using the same objects by computing the absolute difference in valence between participants' affective reactions and their global attitudes and the absolute difference in valence between their cognitive reactions and their overall attitudes.

In the first study, meta-bases were established to be a separate construct from structural bases in their predictive utility for selective information interest. More affective meta-bases predicted a greater proportion of time spent reading an affective message. In the second study, both meta- and structural bases accounted for matching effects such that more affective meta-bases and more affective structural bases predicted greater persuasion by the affective rather than the cognitive message. Finally, in the third study, meta-bases were more likely to predict the use of affect in forming preferences when participants were instructed to be more thoughtful in expressing their attitudes whereas structural bases were more likely to predict the use of affect in forming preferences when people were relatively spontaneous.

The finding from the third study suggests that meta-bases and structural bases influence the motivation and the ability to selectively process information, respectively. When people are deliberative in their evaluations, it is likely that they use information to guide their attitudes in the fashion that they intend to. However, when people are spontaneous, it is likely that they rely on information in their attitudes to the extent that they can efficiently do so. Therefore, affective *meta-bases* participants in the second study might have been more persuaded by the emotions-focused appeal than the beliefs-focused appeal, because they were more motivated to rely on the affective information in their attitude change. Such selective motivation might have stemmed from factors such as the perceived legitimacy of affective information, which might, in turn, have developed from past examples of successful decision making based on affect rather than cognition.

On the other hand, affective *structural* bases participants in the second study might have been more persuaded by the affective than the cognitive appeal because they were more efficient at considering the emotions-related information as they changed their evaluations. This relative ability might have developed from frequent experiences with processing such information, which might, in turn, have been due to the prevalence of such information in one's environment. The notion that structural bases are associated with processing ability is also suggested by previous research demonstrating that for objects that elicit affectively based attitudes, emotional components are more accessible than cognitive components in one's memory (Giner-Sorolla, 2004). The opposite is true for objects that have cognitively based attitudes. It is likely that the higher accessibility of emotions for affective objects, or beliefs for cognitive objects, leads to greater efficiency with which the bases are used to inform attitudes. In summary, the meta-structural distinction is useful in examining matching effects because meta-bases predict processing motivation, whereas structural bases predict processing ability.

MOTIVATION AND ABILITY FACTORS IN ATTITUDES

Motivation and ability factors have been a prevalent feature in attitudes and persuasion models. Both the Elaboration Likelihood Model (ELM) (Petty & Cacioppo, 1986; Petty & Wegener, 1999) and the Heuristic-Systematic Model (HSM) (Chen & Chaiken, 1999) refer to the motivation and ability to process information as important ingredients for an individual to engage in relatively careful scrutiny of a message. If either motivation or ability is low, then such thoughtful processing will be less likely. Processing motivation is influenced by a variety of factors such as the relevance of the attitude object to the individual (Petty & Cacioppo, 1979), the individual's intrinsic inclination for mental challenges (i.e., need for cognition; Cacioppo & Petty, 1982), and the desire for more confidence in one's attitude (Maheswaran & Chaiken, 1991). Processing ability is influenced by variables such as the extent to which an individual is distracted (Petty, Wells, & Brock, 1976). In models of attitudes (e.g., ELM and HSM), the assumption has been that both motivation and ability contribute to information processing and, consequently, attitudes. Although this means that processing motivation and processing ability are viewed as separate contributors to the same phenomenon (i.e., the route of attitude change), the independence of motivation and ability is perhaps more interesting in situations in which one is high and the other is low.

The notion that different consequences are associated with processing motivation and processing ability has been suggested in past research. For instance, relative to individuals who are less confident in their ability to defend their attitudes, individuals who are highly confident in their defensive ability expect themselves to be more successful at generating counter-arguments, and thus, expose themselves to counter-attitudinal information (Albarracin & Mitchell, 2004). However, the same highly confident individuals are not better at counter-arguing than less confident individuals, as evidenced in the former's greater susceptibility to persuasion. In other words, one may be motivated to counter-argue against information that one disagrees with but not be able to actually do so, or vice-versa. Therefore, the motivation to counter-argue might directly influence information exposure, whereas the ability to counter-argue directly impacts attitude resistance.

In the domain of matching effects, the distinction between processing motivation and processing ability was also made in research where the motivation to process information was demonstrated to be a sufficient condition for the advantage of tailoring a strong message to the recipient (See, Petty, & Evans, 2009). This suggests that processing ability is not a necessary condition for matching effects. In this research, the perceived complexity of the presented arguments was matched to the recipient's need for cognition, which refers to one's chronic attraction to mental challenges (Cacioppo & Petty, 1982; see also Cacioppo, Petty, Feinstein, & Jarvis, 1996; Petty, Briñol, Loersch, & McCaslin, 2009; See & Petty, 2007). In Study 2, the results showed that among participants who were high in need for cognition, stronger arguments led to more positive attitudes when the arguments were perceived to be complex but not when they were perceived to be simple. In contrast, among participants who were low in need for cognition, stronger arguments produced more favorable evaluations when the arguments were perceived to be simple but not when they were perceived to be complex. In the study, participants acquired identical background knowledge in order to judge whether the arguments were cogent or not. Therefore, participants' ability for information processing was held constant. Furthermore, the *same* arguments were framed as simple or complex, which means that the arguments placed equal cognitive demands on participants.

Therefore, there is evidence suggesting that processing motivation and processing ability exert independent influences on attitudes. This brings us back to the distinction between meta- and structural bases of attitudes. In recent research, we tested the notion that meta-bases affect attitudes by influencing the motivation for information selection, whereas structural bases do so by influencing efficiency in information selection. Our hypothesis was supported by findings on the differential impact of meta-bases and structural bases on the proportion of time participants spent reading affective information (See, Petty, & Fabrigar, 2010). In this research, individuals with affective meta-bases spent a greater proportion of time reading the affective information than individuals with cognitive meta-bases, thus suggesting that affective meta-bases individuals were more motivated to selectively attend to affective information. In contrast, structural bases predicted selective efficiency such that individuals with affective structural bases spent less reading time to understand the same information as individuals with cognitive structural bases. Various implications arise from such a conceptualization of the meta-structural distinction. For instance, in a situation where affective information on an attitude object is lacking, and thus, the seeking out of such information would be more important for decision making, meta-bases might be more predictive of attitudes and behaviors. However, in a situation where affective information on the object is abundant, and thus, the efficiency in processing such information would be more important, it might be more fruitful to consider structural bases in order to predict attitudes and behaviors. Before discussing the implications of the meta-structural distinction for tailoring information to match attitudes in science education, we describe results from a pilot study that examines whether attitudes toward science subjects have primarily affective or cognitive attitudinal bases.

META- AND STRUCTURAL BASES OF ATTITUDES TOWARD SCIENCE EDUCATION

In a pilot study, we measured the type of attitudinal bases that the three science subjects physics, chemistry, and biology elicit across individuals. We recruited 48 participants (42 females) from the National University of Singapore between the ages of 19 and 24 ($M = 20.5$, $SD = 1.35$). Participants first completed affective, cognitive, and attitudinal items for each of the three subjects on a scale of 1 to 7 (Crites et al., 1994). The affective items required participants to indicate the extent to which their emotions toward physics, chemistry, or biology were negative or positive on eight semantic differential scales (e.g., hateful–loving). The cognitive items asked participants about their beliefs toward each of the subjects on seven semantic differentials (e.g., worthless–valuable). Lastly, participants reported their overall attitudes toward each of the subjects using four semantic differentials (e.g., dislike–like). We computed an absolute affect-attitude discrepancy score and an absolute cognition-attitude discrepancy score for each subject, such that lower affect-attitude discrepancy scores indicate that the attitude object elicits evaluations that are mainly driven by affect, and lower cognition-attitude discrepancy scores indicate that the object elicits evaluations that are mainly driven by cognition.

We also assessed whether each subject was associated with perceptions of greater affective or cognitive reliance in attitudes across participants. The meta-bases for each subject were measured by two questions asking participants to indicate the extent to which their attitudes toward the subject are driven by emotions and beliefs, respectively (See et al., 2008). Participants responded on 7-point scales with endpoint labels of *not at all* and *very much*. Higher scores indicate perceptions that attitudes are very much based on affect and on beliefs, respectively.

In general, participants reported neutral attitudes toward physics, slightly positive attitudes toward chemistry, and even more favorable evaluations toward biology. In addition, participants' neutral attitudes toward Physics were structurally based more on negative affect, despite their positive beliefs about Physics. Participants' slight favorability toward Chemistry was based primarily on their neutral emotions about Chemistry despite their positive beliefs about Chemistry. Finally, participants' positivity towards Biology was based more on their favorable beliefs about Biology than their positive emotions (see Table 6.1).

Of more relevance to the issue of information tailoring, the results indicated that attitudes toward physics were associated with more affective than cognitive structural bases. Attitudes toward chemistry also tended to be more structurally based on affect than cognition ($p < .10$). On the other hand, attitudes toward biology were associated with more cognitive than affective structural bases (see Table 6.2).

TABLE 6.1   Means and Standard Deviations of Emotion, Beliefs, and Global Evaluations toward Each Subject

Subject	Averaged affect		Averaged cognition		Averaged attitude	
	Mean	SD	Mean	SD	Mean	SD
Physics	3.48**	1.19	5.07**	1.21	4.03	1.48
Chemistry	4.12	1.64	5.00**	1.33	4.62*	1.67
Biology	5.07**	1.19	5.84**	.92	5.67**	1.04

Note: Values were tested against mid-point of the 7-point scale. * $p < .05$. ** $p < .01$.

TABLE 6.2   Means and Standard Deviations of Affect-Attitude and Cognition-Attitude Discrepancies toward Each Subject

Subject	Affect-Attitude Discrepancy		Cognition-Attitude Discrepancy	
	Mean	SD	Mean	SD
Physics	.76$_a$	.62	1.27$_b$	1.08
Chemistry	.59$_a$	.58	.79$_a$	.62
Biology	.63$_a$	.56	.40$_b$	.43

Note: Smaller affect-attitude than cognition-attitude discrepancies reflect more affective structural bases. Means in the same row that do not share the same subscript differ at $p < .05$.

TABLE 6.3   Means and Standard Deviations of Perceived Attitudinal Bases toward Each Subject

Subject	Affect		Cognition	
	Mean	SD	Mean	SD
Physics	4.63$_a$	1.95	4.10$_b$	1.73
Chemistry	5.19$_a$	1.61	4.56$_b$	1.80
Biology	5.17$_a$	1.45	5.02$_a$	1.48

Note: Higher scores indicate perceptions that attitudes are based very much on affect and on beliefs, respectively. Means in the same row that do not share the same subscript differ at $p < .05$.

In addition, results demonstrated that physics and chemistry were associated with more affective meta-bases, whereas biology was associated with equally affective and cognitive meta-bases (see Table 6.3). In other words, participants perceived that their evaluations of physics and chemistry were driven more by affect than by cognition, and that their evaluations of biology was based equally on affect and cognition.

As mentioned before, past research has established that meta-bases were more predictive of individuals' reliance of affect in their choice when they were relatively thoughtful in their decision-making, whereas structural bases were more predictive when individuals were relatively spontaneous (See et al., 2008). Therefore, given the results from our pilot study, one implication is that educators could enhance the persuasiveness of their messages by matching them to the meta-bases and structural bases of the attitude objects, depending on whether the intended recipients are relatively deliberative or spontaneous in their decision to pursue a science subject. One could increase students' positivity toward physics using an affective message or strategy (e.g., by generating excitement about studying physics via a teacher's enthusiasm) under both spontaneous and deliberative circumstances, as the structural bases and meta-bases for attitudes physics are affective. The same is true for increasing favorability toward chemistry. On the other hand, given that attitudes toward biology are primarily based on cognition, educators should direct their efforts at persuading students using a cognitive message that highlights the positive characteristics of studying biology, especially if students were relatively spontaneous in their subject choice. However, the advantage of a cognitive message would disappear if students were deliberative, since people perceive their evaluations of biology to be equally driven by affect and cognition.

To test the generalizability of the present findings, future research could use a larger sample size where participants are recruited from various stages of their education (e.g., pre-university). Another limitation that could be addressed by further research is to examine whether the results from our pilot study also apply to male students, since the majority of our participants were female. In addition, given the important role of deliberation in decision making, it would be useful for further research to identify situations in which deliberation is low or high in the education context. For instance, are students more thoughtful when they make decisions about what to pursue in their university education than when they make decisions for their secondary school education?

OTHER MATCHING EFFECTS IN PERSUASION

Functional Bases of Attitudes

Thus far, we have focused on affective and cognitive meta-bases and structural bases. However, there are other ways to consider attitudinal bases. One approach is to examine the functional bases of attitudes. Theorists have proposed that attitudes can serve various purposes such as to express values and standards and to fit in with others (see Katz, 1960; Smith, Bruner, & White, 1956). For instance, female students may hold positive attitudes toward science education because they want to express their personal endorsement of equality or because they want to fit in with their peers who are favorable toward science education. Previous research suggests that the personality variable of self-monitoring may be associated with such value-expressive and social-adjustive attitudinal bases. High self-monitors are individuals who are more concerned about adapting to their social environment, and tend to regulate their behavior to convey positive impressions (Gangestad & Snyder, 2000; Snyder, 1974). On the other hand, low self-monitors are more consistent in their behaviors across situations and tend to look to their internal values and standards to guide their behaviors. Of more relevance, high self-monitors are more persuaded by advertisements that promote the social images that consumers could gain from the use of the product, while low self-monitors were more susceptible to advertisements that emphasized the quality and merits of the product (Snyder & DeBono, 1985). Additional research has demonstrated that matching a message to an individual's self-monitoring tendency is more likely to be effective when the matched message contains strong arguments (Petty & Wegener, 1998). In fact, when the matched message contains weak arguments, the message may not be more effective than the mismatched messages. This suggests that one underlying mechanism for matching effects is increased scrutiny. One question that arises is: Why do people process information that is tailored to them more than information that is not tailored to them?

One approach to addressing the above question is to consider differences in motivation versus ability to process information. As mentioned before, such differences are potentially consequential for predicting attitude change and related behaviors under various circumstances. The greater motivation to process tailored information, relative to non-tailored information, may lead one to seek out tailored information even when it is not widely available in one's surroundings. For instance, female students who hold positive attitudes toward science education because they want to express their endorsement of equality values might be more likely to investigate more specific statistics about gender participation in science education, whereas those who are favorable because they want to fit in with their peers might be more likely to request information on overall statistics about people who choose to major in science subjects. Given their self-selected exposure to the tailored information, attitudes may change or not depending on the extent to which the information is cogent, the information is pro-science education, and the extent to which the individual is able to process the information. The greater efficiency of an individual in processing tailored information than non-tailored information will enable the individual to recognize the merits of cogent information and the flaws of weak information. Consequently, individuals who are presented with strong arguments will be more persuaded than individuals who are presented with weak arguments (cf. Petty & Cacioppo, 1986). Therefore, research on meta- and structural value-expressive versus social-adjustive bases of attitudes would help us predict the effectiveness of various messages by predicting information-seeking tendencies and processing ability, respectively.

Furthermore, research could examine the meta-structural distinction in value-expressive and social-adjustive attitudinal bases at the level of the attitude object (e.g., are attitudes toward physics in general driven by value-expressive bases or social-adjustive bases?) or the level of individual differences (e.g., are female students' attitudes in general driven by value-expressive bases or social-adjustive bases). As mentioned earlier, most research has focused on measuring or manipulating bases at the level of the object or the individual. However, it is possible to also examine attitudinal bases at the level of individual differences regarding a specific attitude object (e.g., are female students' attitudes toward physics driven by value-expressive or social-adjustive functions?). Some preliminary research suggests that examining meta-bases at this level could be useful. In this research, participants reported the extent to which they perceived their attitudes toward video games to be based on their desire to fit in with their peers (See, Fabrigar, & Petty, 2010). Thus, participants were categorized according to whether they had low or high social-adjustive meta-bases in their attitudes toward video games. A week later, participants were randomly assigned to receive either a message that claimed that the majority of undergraduates did not approve of video games as an educational tool or a control message that also argued against educational video games. Participants then reported how much they would pay for educational video games. Therefore, lower amounts indicated greater persuasion. Results showed that high social-adjustive meta-bases participants were willing to pay less money for educational video games after they read the social message than after they read the control message. In contrast, low social-adjustive meta-bases participants were willing to pay less money for educational video games after the control message than after the social message. Therefore, this research suggests that examining social-adjustive meta-bases at the level of individual differences regarding a specific attitude object (i.e., educational video games) is informative for predicting matching effects.

Social Identity as an Attitudinal Basis

Researchers have also examined attitudinal bases by considering the recipient's group membership. One difference in such matching effects is that instead of matching the contents or framing of the message to the recipient, information about the *source* of the information is made available that matches or mismatches the recipient's group membership. Previous research has found that when the source of the message comes from an ingroup, persuasion is more likely to occur than when the source comes from an outgroup (e.g., Fleming & Petty, 1997; Mackie, Worth, & Asuncion, 1990). In addition, such a matching effect is more likely when the attitude object is irrelevant or mundane or when the attitude object is relevant but the information presented is cogent (Mackie et al., 1990), and when the recipient is highly identified with the group (Fleming & Petty, 1997). Therefore, previous research suggests that individuals may have attitudes that are perceived to be based on their group membership or structurally driven by their group membership.

Applied to science education attitudes, especially among those female students who are disinterested in science subjects and highly identified with their gender, having a female teacher encourage them to pursue science education might be more effective than having a male teacher do so. In addition, the conceptualization of meta-bases could enhance our understanding of how such matching works. Using the same example, female students who perceive that their lack of interest in science education is due to their gender identity might be more interested in information from a female teacher rather than a male teacher because they expect that the former would present more legitimate or more relevant arguments. Consequently, compared to female students who do not perceive that their disinterest in science is due to their gender identity, female students who are high in gender meta-bases would be more motivated to process arguments from a female teacher than a male teacher. This is consistent with previous research that suggests that increased processing is one underlying mechanism for group membership matching effects. As mentioned earlier, in some research on group membership matching effects, more attitude change occurred when the information came from an ingroup source than an outgroup source for a topic that was group-relevant, but only when the information was cogent (Mackie et al., 1990).

What about group membership *structural* bases? The group membership matching effects examined in most of the literature involve tailoring the message source and not the message contents to the recipient. This means that the information presented is substantively identical in all conditions. Therefore, it is less likely that group membership effects that involve structural bases are due to increased processing ability. Returning to the example above, female students who are high in gender structural bases would not be processing information from a female teacher more than a male teacher because they were somehow more efficient at processing the information from the female teacher. However, it is possible that group membership structural bases predict persuasion when the message source serves as a peripheral cue (cf. Petty & Cacioppo, 1986; see also Fleming & Petty, 2000). In other words, the heuristic "Being female is important to me so I should listen to a recommendation from another female" may be more accessible for female students whose attitudes are structurally based on their female group membership than those whose attitudes are not. This prediction is consistent with previous research showing that structural bases are more likely to impact selective information reliance when individuals are being relatively spontaneous in making choices (See et al., 2008).

Self-Schema as an Attitudinal Basis

While an individual can view him- or herself as part of a group, an individual can also have perceptions of specific traits or attributes that he or she possesses. Self-schemas direct attention to information and guide behaviors (Markus & Wurf, 1987). Indeed, previous research has demonstrated that tailoring the contents or the frame of a message to the recipient increases persuasion especially when the arguments are strong. For instance, participants with a religious self-schema found arguments emphasizing the sacredness of life more convincing than arguments that focused on the right to life, whereas the opposite pattern was observed among participants with a legalistic self-schema (Cacioppo, Petty, & Sidera, 1982). As with research on functional and group membership bases of attitudes, recent studies have also demonstrated that such self-schema matching effects could be due to increased processing (Wheeler, Petty, & Bizer, 2005). In the research, participants allocated more attention and effort to scrutinizing the advertisement that was framed to their extroversion or introversion (Study 1) as well as their need for cognition (Study 2), and therefore, were more persuaded by the matched information when it was cogent.

It is worth slightly digressing here to note that need for cognition is conceptually and empirically different from cognitive meta-bases. As mentioned earlier, individuals who are high in need for cognition are intrinsically motivated to process complex information. Such information could be affective and cognitive in nature. On the other hand, cognitive meta-bases individuals are more willing to rely on cognition in their attitudes, which suggests that (1) they process cognitive information for the sake of forming preferences (and not for the sake of cognitive processing itself), and (2) they process cognitive information regardless of whether it is simple or complex. Need for cognition is also distinguishable from cognitive structural bases, as the latter construct reflects the ability (rather than the motivation) to process cognitive information and use it in forming preferences. Indeed, in previous research where need for cognition was measured within the same study as attitudinal bases, no correlation was found between need for cognition and affective or cognitive attitudinal bases (Huskinson & Haddock, 2004; See et al., 2008).

In summary, information that is tailored to be consistent with the recipient's self-schema is more persuasive than information that is not. In addition, the advantage of tailored information over non-tailored information is more likely to hold when both are strong arguments. This is especially true when the recipient is moderate in his or her processing tendencies (cf. Petty & Cacioppo, 1986; Petty & Wegener, 1999). Furthermore, the increased scrutiny could be due to both the motivation and ability to process information. When processing tendencies are moderate, both meta- and structural bases might increase the extent of scrutiny of the message. For example, students whose self-schemas involve being creative might perceive that their negativity toward studying physics is based on their creative personality. Such students would be more interested in scrutinizing a message that discusses the positive relationship between studying physics and enhancing creativity than their counterparts whose self-schemas also include being creative but whose physics attitudes are not perceived to be based on their creativity. Consequently, the former group would change their attitudes toward studying physics as a function of the strength of the message, whereas the latter group would not. Such a pattern is also likely to occur when people are under relatively high processing tendencies—for instance, if their decision to study physics is going to impact them next year rather than three years later (see Petty & Cacioppo, 1979). This prediction is consistent with previous research, which demonstrated that meta-bases are more likely to be impactful when people are moderately or highly deliberative in their decision-making (See et al., 2008).

Structural bases are also likely to be influential under moderate deliberation. However, compared to meta-bases, structural bases are more likely to impact attitudes and behaviors when people are relatively spontaneous rather than highly deliberative in their decision making (See et al., 2008). Using the same example, if the decision to study physics is going to impact a student three years rather than a year later, then the student might process the tailored message more than the non-tailored message to the extent that he or she is more efficient (rather than motivated) at doing so. Such efficiency might depend on the degree of association between self-schema and physics attitude in one's memory, which might have developed from frequent exposure to such an association at home or at school.

Matching to Regulatory Orientation

Another active research area that examines matching effects consists of studies investigating an individual's regulatory orientation. According to Regulatory Focus Theory (Higgins, 1997), differences in the manner of goal pursuit might stem from situational and personality factors. A promotion focus is associated with goal pursuit in terms of ideals, and entails an orientation toward accomplishment as well as sensitivity to the presence and absence of gains; a prevention focus is associated with goal pursuit in terms of responsibilities, and entails an orientation toward security as well as sensitivity to the presence and absence of losses. Past research has demonstrated that matching the message to one's promotion focus (i.e., by highlighting opportunities) or to one's prevention focus (i.e., by highlighting safety) leads to greater or less persuasion (Cesario, Grant, & Higgins, 2004). Other research has also shown similar matching effects by using self-views as a proxy for regulatory focus (Aaker & Lee, 2001). In this research, participants had promotion or prevention goals as a function of either the situational prime or their cultural background. Individuals from individualistic cultures (i.e., Americans) were proposed to have promotion goals, whereas collectivistic cultures (i.e., Chinese) were proposed to have prevention goals. Of more relevance, attitudes were more positive when the message was matched to the recipient's regulatory orientation than when it was mismatched. As with other matching effects, this pattern was more likely when the matched message was strong rather than weak. Therefore, the overall evidence suggests that individuals may base their evaluations on their regulatory orientation.

Further research could examine the extent to which students are not very favorable toward science education because they think that such education does not provide opportunities for career advancement or because they do not think that such education leads to safe career options. Tailored messages could directly target such concerns to promote positivity toward science education. Moreover, it would be useful to examine how the dominance of one type of concerns over the other develops, whether in one's home, school, or culture. For example, in the school environment, it is possible that teachers' expectations might induce students to learn to work toward accomplishing the best grade they could attain (i.e., promotion-focus) or to avoid obtaining a failing grade (i.e., prevention focus; see Cesario et al., 2004). The regulatory orientation then becomes a basis for students' attitudes toward science education. Such a development is probably likely for *structural* bases that involve regulatory orientation. On the other hand, students may form promotion- or prevention-focus meta-bases due to perceived efficacy in fulfilling ideals or responsibilities. Importantly, as found in research on affective and cognitive meta-bases and structural bases (See et al., 2008), perceptions of selective reliance may not correspond to actual reliance. Therefore, both meta- and structural bases would have to be considered in order to predict susceptibility to tailored information.

CONCLUSION

In this chapter, we have discussed different types of matching effects. These various matching effects are especially relevant to promoting positivity toward science education depending on how a researcher wishes to apply them. For example, researchers who are interested in developing teaching strategies in the classroom that promote interest in science subjects (e.g., by creating a warm and supportive environment; see Osborne et al., 2003) might benefit from examining the affective and cognitive meta-bases and structural bases of science education attitudes. On the other hand, researchers who are interested in how parents and peers influence science education attitudes might find it more useful to examine the social-adjustive meta- and structural bases of science education attitudes. In fact, there are other variables in matching effects research that could be relevant to science education attitudes. One such example is the message recipient's doubt (e.g., Briñol, Petty, & Wheeler, 2006) or confidence (Tormala, Rucker, & Seger, 2008).

Furthermore, there are various ways in which information or other aspects of a persuasive context (e.g., the message frame, the message source, etc.) could be tailored to the message recipient in order to promote attitude change. For instance, although we focused on explaining why information tailoring works for functional bases of attitudes, it is certainly possible to match the message *source* to one's functional basis (e.g., Ziegler, von Schwichow, & Diehl, 2005). We propose that when the information itself is matched to the recipient, persuasion could be impacted by processing motivation or processing ability, or both. However, when the information remains the same but only its frame or its source is tailored to the recipient, persuasion is more likely to be influenced by processing motivation than processing ability.

As illustrated by the various examples throughout the chapter, understanding the mechanisms for matching effects will enhance our ability to predict consequences in various circumstances. In conclusion, we hope that the meta-structural distinction will prove to be a fruitful and generative approach when designing messages to promote positivity toward science education.

REFERENCES

Aaker, J.L., & Lee, A.Y. (2001). "I" seek pleasures and "we" avoid pains: The role of self-regulatory goals in information processing and persuasion. *Journal of Consumer Research, 28,* 33–49.

Albarracin, D., & Mitchell, A. L. (2004). The role of defensive confidence in preference for proatttitudinal information: How believing that one is strong can sometimes be a defensive weakness. *Personality and Social Psychology Bulletin, 30,* 1565–1584.

Bassili, J.N. (1996). Meta-judgmental versus operative indexes of psychological attributes: The case of attitude strength. *Journal of Personality and Social Psychology, 71,* 637–653.

Breckler, S.J. (1984). Empirical validation of affect, behavior, and cognition as distinct components of attitude. *Journal of Personality and Social Psychology, 47,* 1191–1205.

Briñol, P., Petty, R. E., & Wheeler, S. C. (2006). Discrepancies between explicit and implicit self-concepts: Consequences for information processing. *Journal of Personality and Social Psychology, 91,* 154–170.

Cacioppo, J.T. & Petty, R.E. (1982). The need for cognition. *Journal of Personality and Social Psychology, 42,* 116–131.

Cacioppo, J.T., Petty, R.E., Feinstein, J.A., & Jarvis, W.B.G. (1996). Dispositional differences in cognitive motivation: The life and times of individuals varying in Need for Cognition. *Psychological Review, 119,* 197–253.

Cacioppo, J. T., Petty, R. E., & Sidera, J. A. (1982). The effects of a salient self-schema on the evaluation of proattitudinal editorials: Top-down versus bottom-up message processing. *Journal of Experimental Social Psychology, 18,* 324–338.

Cesario, J., Grant, H., & Higgins, T.E. (2004). Regulatory fit and persuasion: Transfer from "Feeling Right." *Journal of Personality and Social Psychology, 86,* 388–404.

Chen, S., & Chaiken, S. (1999). The heuristic-systematic model in its broader context. In S. Chaiken & Y. Trope (Eds.), *Dual-process theories in social and cognitive psychology* (pp. 73–96). New York: Guilford Press.

Crites, S.L., Fabrigar, L.R. & Petty, R.E. (1994). Measuring the affective and cognitive properties of attitudes: Conceptual and methodological issues. *Personality and Social Psychology Bulletin, 20,* 619–634.

Davidson, A.R., Yantis, S., Norwood, M., & Montano, D.E. (1985). Amount of information about the attitude object and attitude-behavior consistency. *Journal of Personality and Social Psychology, 49,* 1184–1198.

Eagly, A. H., & Chaiken, S. (1993). *The psychology of attitudes.* Fort Worth, TX: Harcourt Brace Jovanovich.

Eagly, A.H., Mladinic, A., & Otto, S. (1994). Cognitive and affective bases of attitudes toward social groups and social policies. *Journal of Experimental Social Psychology, 30,* 113–137.

Esses, V.M., Haddock, G., & Zanna, M.P. (1993) Values, stereotypes, and emotions as determinants of intergroup attitudes. In D.M. Mackie & D.L. Hamilton (Eds.), *Affect, cognition, and stereotyping: Interactive processes in group perception* (pp. 137–166). San Diego, CA: Academic Press.

Fabrigar, L.R., & Petty, R.E. (1999). The role of affective and cognitive bases of attitudes in susceptibility to affectively and cognitively based persuasion. *Personality and Social Psychology Bulletin, 25,* 363–381.

Fishbein, M., & Ajzen, I. (1975). *Belief, attitude, intention, and behavior: An introduction to theory and research.* Reading, MA: Addison-Wesley.

Fleming, M. A., & Petty, R. E. (1997). Social identity with an ingroup moderates the effectiveness of an ingroup source as a persuasive cue. Unpublished raw data. Ohio State University, Columbus.

Fleming, M.A., & Petty, R.E. (2000) Identity and persuasion: An elaboration likelihood approach. In D.J. Terry & M.A. Hogg (Eds.), *Attitudes, behavior, and social context: The role of norms and group membership (pp. 171–199).* Mahwah, NJ: Erlbaum.

Gangestad, S. W., & Snyder, M. (2000). Self-monitoring: Appraisal and reappraisal. *Psychological Bulletin, 126,* 530–555.

Gawronski, B. (2007). Attitudes can be measured! But what is an attitude? [Special issue]. *Social Cognition, 25*(5).

Giner-Sorolla, R. (2004). Is affective material in attitudes more accessible than cognitive material? The moderating role of attitude basis. *European Journal of Social Psychology, 34,* 761–780.

Higgins, E. T. (1997). Beyond pleasure and pain. *American Psychologist, 52,* 1280–1300.

Huskinson, T.L.H., & Haddock, G. (2004). Individual differences in attitude structure: Variance in chronic reliance on affective and cognitive information. *Journal of Experimental Social Psychology, 40,* 82–90.

Katz, D. (1960). The functional approach to the study of attitudes. *Public Opinion Quarterly, 24,* 163–204.

Katz, D., & Stotland, E. (1959). A preliminary statement to a theory of attitude structure and change. In S. Koch (Ed.), *Psychology: A study of a science (Vol. 3): Formulations of the person and the social context* (pp. 423–475). New York: McGraw-Hill.

Krosnick, J. A., Boninger, D. S., Chuang, Y. C., Berent, M. K., & Carnot, C. G. (1993). Attitude strength: One construct or many related constructs? *Journal of Personality and Social Psychology, 65,* 1132–1151.

Krosnick, J. A., & Petty, R. E. (1995). Attitude strength: An overview. In R. E. Petty, and J. A. Krosnick (Eds.), *Attitude strength: Antecedents and consequences* (pp. 1–24). Hillsdale, NJ: Erlbaum.

Mackie, D. M., Worth, L. T., & Asuncion, A. G. (1990). Processing of persuasive in-group messages. *Journal of Personality and Social Psychology, 58,* 812–822.

Maheswaran, D. & Chaiken, S. (1991). Promoting systematic processing in low-motivation settings: Effect of incongruent information on processing and judgment. *Journal of Personality and Social Psychology, 61,* 13–25.

Markus, H., & Wurf, E. (1987). The dynamic self-concept: A social psychological perspective. In M. R. Rosenzweig & L.W. Porter (Eds.), *Annual Review of Psychology, 38,* 299–337.

Osborne, J., Simon, S., & Collins, S. (2003). Attitudes towards science: A review of the literature and its implications. *International Journal of Science Education, 25,* 1049–1079.

Petty, R. E., Briñol, P., Loersch, C., & McCaslin, M. J. (2009). The need for cognition. In M. R. Leary & R. H. Hoyle (Eds.), *Handbook of individual differences in social behavior* (pp. 318–329). New York: Guilford Press.

Petty, R. E., & Cacioppo, J. T. (1979). Issue involvement can increase or decrease persuasion by enhancing message-relevant cognitive processes. *Journal of Personality and Social Psychology, 37,* 1915–1926.

Petty, R.E. & Cacioppo, J.T. (1986). The elaboration likelihood model of persuasion. In L. Berkowitz (Ed.), *Advances in experimental social psychology* (Vol. 19, pp. 123–205.) New York: Academic Press.

Petty, R. E., & Wegener, D. T. (1998). Matching versus mismatching attitude functions: Implications for scrutiny of persuasive messages. *Personality and Social Psychology Bulletin, 24,* 227–240.

Petty, R. E., & Wegener, D. T. (1999). The elaboration likelihood model: Current status and controversies. In S. Chaiken & Y. Trope (Eds.), *Dual process theories in social psychology* (pp. 41–72). New York: Guilford Press.

Petty, R. E., Wells, G. L., & Brock, T. C. (1976). Distraction can enhance or reduce yielding to propaganda: Thought disruption versus effort justification. *Journal of Personality and Social Psychology, 34,* 874–884.

Petty, R. E., Wheeler, S.C., & Bizer, G. (2000). Matching effects in persuasion: An elaboration likelihood analysis. In G. Maio & J. Olson (Eds.), *Why we evaluate: Functions of attitudes* (pp. 133-162). Mahwah, NJ: Erlbaum.

Priester, J.R., & Petty, R.E. (1996). The gradual threshold model of ambivalence: Relating the positive and negative bases of attitudes to subjective ambivalence. *Journal of Personality and Social Psychology, 71,* 431–449.

Rosenberg, M.J., & Hovland, C.I. (1960). Cognitive, affective and behavioral components of attitudes. In M.J. Rosenberg, C.I. Hovland, G.J. McGuire, R.P. Abelson, & J.W. Brehm (Eds.), *Attitude organization and change* (pp. 1–14). New Haven, CT: Yale University Press.

See, Y.H.M., Fabrigar, L.R., & Petty, R.E. (2010, May). *Matching a message to chronic social-adjustive meta-bases increases persuasion.* Poster presented at the American Psychological Society Annual Convention, Boston, MA.

See, Y. H. M., & Petty, R. E. (2007). The need for cognition. In R. Baumeister & K. Vohs (Eds), *Encyclopedia of social psychology* (Vol. 2, pp. 611–613). Thousand Oaks, CA: Sage Publications.

See, Y.H.M., Petty, R.E., & Evans, L.M. (2009). The impact of perceived message complexity and need for cognition on information processing and attitudes. *Journal of Research in Personality, 43,* 880–889.

See, Y.H.M., Petty, R.E., & Fabrigar, L.R. (2008). Affective and cognitive meta-bases of attitudes: Unique effects on information interest and persuasion. *Journal of Personality and Social Psychology, 94,* 938–955.

See, Y.H.M., Petty, R.E., & Fabrigar, L.R. (2010, January). *Meta- and structural bases differentially predict selective processing.* Poster presented at 11th annual Society for Personality and Social Psychology meeting, Las Vegas, NV.

Simpson, R.D., & Oliver, J. S. (1990). A summary of major influences on attitude toward and achievement in science among adolescent students. *Science Education, 74,* 1–18.

Smith, M. B., Bruner, J. S., & White, R. W. (1956). *Opinions and personality.* New York: Wiley.

Snyder, M. (1974). Self-monitoring of expressive behavior. *Journal of Personality and Social Psychology, 30,* 526–537.

Snyder, M., & DeBono, K.G. (1985). Appeals to image and claims about quality: Understanding the psychology of advertising. *Journal of Personality and Social Psychology, 49,* 586–597.

Thompson, M.M., Zanna, M.P., & Griffin, D.W. (1995). Let's not be indifferent about (attitudinal) ambivalence. In R. E. Petty & J. A. Krosnick (Eds.), *Attitude strength: Antecedents and consequences* (pp. 361–386). Mahwah, NJ: Erlbaum.

Tormala, Z.L., Rucker, D.D., & Seger, C.R. (2008). When increased confidence yields increased thought: A confidence-matching hypothesis. *Journal of Experimental Social Psychology, 44,* 141–147.

Wheeler, C.S., Petty, R.E., & Bizer, G.Y. (2005). Self-schema matching and attitude change: Situational and dispositional determinants of message elaboration. *Journal of Consumer Research, 31,* 787–797.

Wood, W., Rhodes, N., & Biek, M. (1995). Working knowledge and attitude strength: An information-processing analysis. In R. E. Petty & J. A. Krosnick (Eds.), *Attitude strength: Antecedents and consequences* (pp. 283–313). Mahwah, NJ: Erlbaum.

Zanna, M.P. & Rempel, J.K. (1988). Attitudes: A new look at an old concept. In D. Bar-Tal & A.W. Krugranski (Eds.), *The social psychology of knowledge* (pp. 315–334). New York: Cambridge University Press.

Ziegler, R., von Schwichow, A., & Diehl, M. (2005). Matching the message source to attitude functions: Implications for biased processing. *Journal of Experimental Social Psychology, 41,* 645–653.

AUTHOR NOTE

Ya Hui Michelle See and Bernice L.Z. Khoo, Department of Psychology, National University of Singapore.

Preparation of this chapter was supported by grants from National Research Foundation NRF2008IDM-IDM001-028 and Ministry of Education (MOE) Academic Research Fund R581-000-079-112.

Correspondence regarding this chapter should be addressed to Ya Hui Michelle See, Department of Psychology, National University of Singapore, 9 Arts Link, Singapore 117570. E-mail: psysyhm@nus.edu.sg

CHAPTER 7

ASSESSMENT PRACTICES FOR UNDERSTANDING SCIENCE-RELATED ATTITUDES

Carina M. Rebello, Stephen B. Witzig, Marcelle A. Siegel, and Sharyn K. Freyermuth
University of Missouri

ABSTRACT

This chapter focuses on ways in which attitudes are assessed. There is a wide diversity in attitude assessment practices; an appropriate way to examine these practices is through the lens of the COI (cognition, observation, and interpretation) assessment triangle as reported in *Knowing What Students Know* (Pellegrino, Chudowsky, & Glaser, 2001). The usage of the COI assessment triangle provides a unique approach to analyzing current attitude-based assessments and assessment design. We provide an overview of existing research methods and instruments used to assess attitudes followed by a brief discussion about the meaning of "attitudes towards science." We focus the COI analysis on two articles and discuss assessment strategies, assessment design, ways in which results are interpreted, and implications for assessment practices in the future.

Attitude Research in Science Education, pages 199–218

INTRODUCTION

Attitudes can be learned or influenced from a variety of experiences and situations (George, 2000). Attitudes are also an integral factor in regard to student success in science and choosing to pursue science (Bennett, 2001). Thus educators, education researchers, and policy makers have a critical interest in understanding students' attitudes in relation to science interest and understanding, as well as attitudes towards the learning of science (Blalock, Lichtenstein, Owen, Pruski, Marshall, & Toepperwein, 2008). To date there has been considerable research into science-related attitudes (for a review, see Gardner, 1975; Osborne, Simon, & Collins, 2003; Reid, 2006; Schibeci, 1984). This effort has largely contributed to the desire to increase student interest, performance, and science retention, along with improving science literacy (Siegel & Ranney, 2003).

Assessments can provide a means for measuring and monitoring educational and achievement outcomes of students, assist learning, and to evaluate programs (Pellegrino, Chudowsky, & Glaser, 2001). By assessing science-related attitudes, educators can obtain both formative and summative feedback about their students and teaching. These results can allow educators to make better informed decisions with regard to their curriculum and instruction, program evaluation, and accountability purposes. Research has revealed that effective instruction can potentially improve students' attitudes towards science (Koballa & Glynn, 2007; Siegel & Ranney, 2003). Thus, assessments are viewed as an instrumental factor in education and instruction. Yet assessments can only be effective and meaningful if constructed through an operative assessment design framework. Little research exists, however, regarding the construction of science-related attitude instruments adhering to current assessment design practices.

A review of previous attitudes research shows that there have been a considerable number of attitude assessments constructed (Osborne et al., 2003). Yet most of these attitude assessments generally focus on exploring students' beliefs, values, images of scientists, or the affective orientation toward science, and other factors associated with attitudes (Siegel & Ranney, 2003). Less discussion has occurred with regard to the application of these assessments as classroom or large-scale assessments or applying these assessments to program evaluation. There are several interesting questions that can be raised regarding this issue, such as: Should teachers use an attitude instrument in their classroom? What type of instrument should teachers use (observations, interviews, survey, etc.)? How should an attitude instrument be used in teaching? Would a pre/post attitude instrument, if designed correctly, be effective in program evaluation? Additionally, does the instrument actually measure an individual's attitudes? Or, is the reporting of the results fair? In order for researchers to address these issues, several considerations should be made about the instrument itself, specifically with regard to the design of the instrument.

In order to examine the effectiveness and usefulness of an attitude instrument, we need to first examine the design of the instrument through the lens of an assessment design framework. If meaningful inferences are to be made from attitude assessments, these assessments ideally should be designed such that they reflect reasoning from evidence. A framework for assessment design that addresses assessment characteristics such that the assessment reflects reasoning from evidence is the assessment triangle (Pellegrino et al., 2001). It is argued that effective assessments are developed if the essential characteristics (cognition, observation, and interpretation) of the assessment triangle (COI framework) are applied to their design.

FRAMEWORK—ASSESSMENT TRIANGLE

At this present time, assessments have been strongly influenced by challenging academic standards in order to heighten educational quality (Pellegrino et al., 2001). A recent U.S. National Research Council Committee (Pellegrino et al., 2001) report, charged with the purpose of evaluating and improving assessments, emphasized that assessments should be defined as a process of reasoning from evidence, portrayed as a system of three interconnected element—cognition, observation, and interpretation (see Figure 7.1). This includes "a model of how students represent knowledge, task or situations that allow one to observe students' performance, and an interpretation method for drawing inferences from the performance evidence thus obtained" (Pellegrino et al., 2001, p. 2).

Each element underlies all assessments and comprises the assessment triangle (COI framework). The triangle represents the components that

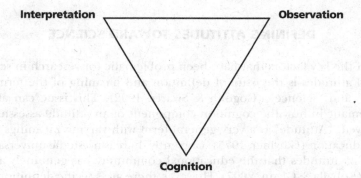

Figure 7.1   The assessment triangle. (Pellegrino et al., 2001. *Knowing What Students Know* (p. 44). Washington, D.C.: National Academy Press)

are essential for designing and implementing assessments (Tomanek, Talanquer, & Novodvorsky, 2008). A recommendation brought forth by the council, in order to achieve effective and meaningful assessments, is that assessments should be explicitly designed and connected to each of the three elements as a coordinated whole (Pellegrino el al., 2001). Assessments should consist of a model of cognition and learning that best reflects how students represent knowledge and develop competence (cognition). Tasks or situations should be used within the assessments to allow for the collection of evidence about students' knowledge and skills. These tasks should be carefully linked to the cognitive model of learning and support the kinds of inferences made based upon results obtained (observation). Finally, the assessment should support a method for drawing inferences from observations (interpretation) (Pellegrino et al., 2001).

Assessments are effective and useful only to the degree in which the cognition, observation, and interpretation components of the assessment are in synchrony (Pellegrino et al., 2001). If assessment tasks are designed without clearly considering whether and how those tasks require students to demonstrate the targeted knowledge and skills, the data collected from such an assessment will potentially not address the intended goals of the assessment. Also, if tasks were designed without thinking about how to analyze the performance appropriately, the data may not allow for the detection of students' strengths or weaknesses. Therefore, each component of the assessment triangle within the assessment needs to support the kinds of inferences educators now want to draw about student attitude and achievement (Pellegrino et al., 2001). This will be especially important if we are to determine whether an attitude instrument actually measures an individual's attitudes within its intended use, and if results are fairly and accurately interpreted such that informed decisions can be made to further benefit learning.

DEFINING ATTITUDES TOWARD SCIENCE

One of the key factors that have been problematic for research in science-related attitudes is the issue of definition and meaning of the terms "attitude". and "science" (Gogolin & Swartz, 1992). This issue can also be problematic in how the cognition component of an attitude assessment is perceived. "Attitude" is a very general term with varying meanings in science education (Gardner, 1975). Currently there is no single universal definition of attitudes that the educational community has genuinely agreed upon (Koballa & Glynn, 2007). However, there are specific definitions that appear in literature (e.g., Ramsden, 1998; Schibeci, 1984; Shrigley, Koballa, & Simpson, 1988, in Koballa & Glynn, 2007;). Kind, Jones, and Barmby

(2007) describe a common definition of attitude as consisting of three inter-linked components: cognition, affect, and behavior.

In science education, science-related attitudes have been broken into two broad categories: *attitudes towards science* and *scientific attitudes*. Attitudes towards science encompass such things as interest in science, the value of science, attitudes about a specific science domain, attitudes towards a science issue, or attitudes towards social responsibility in science (Gardner, 1975; Schibeci, 1984). There are a range of components that researchers have incorporated into their measures (e.g., Osborne et al., 2003). Generally there is a distinct attitude object (science, scientist, science lesson, laboratory experiments, etc.) to which the individual responds favorably or unfavorably (Gardner, 1975). A more specific definition of attitude that applies to the previous idea is stated in Kind and colleagues' (2007) article as "the feelings that a person has about an object, based on their beliefs about that object" (p. 873). By applying this definition, "an attitude towards science measure can be thought of as mapping students' cognitive and emotional opinions about various aspects of science" (Kind et al., 2007, p. 873). Scientific attitudes, on the other hand, consist of traits or styles of thinking as a scientist. These traits can include open-mindedness, the need for evidence and verification, objectivity, applying a questioning approach, honesty, and scepticism (Blalock et al., 2008; Gardner, 1975; Osborne et al., 2003). Essentially, scientific attitudes include traits that characterize scientific thinking (Osborne, et al., 2003).

This summary of attitudes represents fairly traditional views of what attitudes are and what should be measured. In our discussion we will look beyond these perspectives and propose additional perspectives that future research should consider. How attitude assessment developers define attitudes will be important in how the assessment instrument is effectively designed and implemented.

COMMON ATTITUDE MEASURES

Review of Existing Measures

Traditionally, the assessment of attitudes measures an individual's expressed preference or feeling towards an object (science, scientists, etc.) (Osborne, et al., 2003). There are a variety of approaches to assessing science-related attitudes. These generally are self-report instruments providing quantitative measures of attitudes, particularly Likert-scale questionnaires (Koballa & Glynn, 2007). Instruments to assess attitudes typically include preference ranking, attitude scales (e.g., Thurstone and Likert scale items), interest inventories, subject enrollment, and qualitative methods (e.g., student drawing, interview, and physiological expressions) (Kind et al., 2007; Koballa & Glynn, 2007).

Each assessment method presents its own unique disadvantage(s) in order to provide an accurate and valid picture of an individual's attitudes to a particular aspect of science (Reid, 2006). It has been well documented that attitude instruments are weakened by the lack of a clearly defined concept of attitude and having weak psychometric quality. Having a weak psychometric quality refers to lacking internal consistency, construct validity, and unidimensionality (Kind et al., 2007). Another concern for attitude instrument development is the lack of establishing a theoretical or conceptual basis (Blalock et al., 2008).

Extensive work has critically analyzed science-related attitude instruments' definitions of attitudes, validity and reliability reports, and use of theoretical basis (e.g., Blalock et al., 2008; Gardner, 1975; Kind et al., 2008; Munby, 1997; Osborne et al., 2003; Ramsden, 1998). Yet there has been no examination of attitude assessments design through a lens of an assessment development framework. Although the above critiques are important for the development and utilization of an attitudes assessment, assessments also need to be grounded and cohesively aligned with three key features of assessment design—cognition, observation, and interpretation. This framework for developing assessments can help ensure fair reporting and interpretation of results. Pellegrino and colleagues (2001) argue that good assessment design should include and interconnect the essential features of assessment—a model of cognition and learning, an interpretation model, and carefully developed assessment tasks that match the interpretation of evidence model and elicits the knowledge and cognitive process that the model of learning suggests. By providing a model of cognition and learning, you can gain a developmental perspective, showing ways in which learners progress. If assessments lack a model of cognition and learning, any interpretation technique applied will produce limited information about progress. Conversely, if a model based on a modern understanding of how students learn is included in the assessment design, but sufficient tools to interpret the data are lacking; again the assessment will produce limited information (Pellegrino et al., 2001).

Review of Attitude Assessments

The goal of this assessment review is to identify and evaluate science attitude instruments based upon their design in relation to the assessment triangle as proposed by Pellegrino and colleagues (2001). A search was conducted on existing academic databases using a variety of search terms and names of authors known to publish on the topic of science attitudes and measurement.

In addition, reference lists of identified works were examined. Criteria were then developed to select two instruments for further review and analysis. In the next section, we present our review and critiques of the works by Gogolin and Swartz (1992) and Osborne and Collins (2000, 2001).

ANALYSIS OF ATTITUDE ASSESSMENTS

Next, we frame the instrument around each component of the assessment triangle (cognition, observation, and interpretation) while describing the instrument and the study involved in the development of the instrument. We will also discuss the connections between each of the components in the assessment triangle in regards to the instrument and offer a critique that may improve the effectiveness of the instrument.

Osborne & Collins (2000, 2001)

Context

Osborne & Collins (2000) conducted a large-scale study in London, Leeds, and Birmingham with 16-year-old, year 11 students, their parents, and teachers. The goal of the study was to explore the range of views, comments, and attitudes that students and their parents have regarding the school science curriculum, the aspects they found interesting and/or valuable, and its future content. They were also interested in investigating teachers' responses to those views. More specifically, Osborne and Collins's (2001) research aim was to determine views, of both students and parents, on "the kind of scientific knowledge, skills, or understanding that they need for dealing with everyday life; The aspects of science that they find interesting; The value of the content of the school science education that they received; and The future content of the science curriculum for all" (p. 443).

The study was driven by the view that their current state of the school science curriculum may be in need of revision. The authors suggest that it is important to determine the various aspects that students and their parents value and use in their everyday lives. Osborne and Collins (2000, 2001) further argue that all people commonly engage with science and technology in a broad range of contexts (e.g., health nutrition, pollution, etc.) and because of these experiences it is essential to articulate their views on the debate of curriculum reform. Thus, participants were selected in order to represent a cross-section of the population to ensure that as many views as possible were gathered and that the data were a comprehensive representation of the population. A qualitative study was conducted in order to explore students' in-depth view and rationale (Osborne & Collins, 2000, 2001).

Cognition

Based on the scope and aims of Osborne and Collins's (2000) study and their application of focus groups, it appears, although not explicitly, that they take a social stance on a model of cognition and learning, thus addressing the cognition portion of the assessment triangle. From their research report, we infer that they view attitude as an unstable, dynamic construct that is influenced by various social experiences and should be probed in order to be analyzed. In order to elicit and assess these attitudes, they employ group interaction—a naturalized social setting in which participants may challenge those holding varying views to elaborate and provide justification. The use of focus group interviews allows for the authors' goal of exploring attitudes, beliefs, and values that are commonly held with the populace. Therefore, focus groups appear to be a viable option for their study, creating a social dynamic context in which attitudes, beliefs, and values can be extracted. However, an important critique in developing an effective and meaningful assessment is having a clear, explicit model of cognition and learning based on empirical knowledge of how individuals learn in a given domain (Pellegrino et al., 2001). The authors' discussion around the development of their assessment does not appear to provide such a model clearly and explicitly.

Observation

The principal methodology adopted by Osborne and Collins (2000) to elicit views was focus group interviews with groups of students, parents, and teachers. Interview protocols for each group were developed to elicit attitudes, beliefs, and values that target the authors' aim of the research. The appeal of focus groups is that they can provide a dynamic, yet non-threatening social setting in which to probe and challenge the participants' views and positions regarding the issues.

Osborne and Collins (2000) provide an explanatory contrast of their chosen methodology with individual interviews and questionnaire. Although individual interviews can allow for flexibility and the potential to follow up points that emerge from responses, they can also promote situations in which respondents will provide answers that they think the interviewer wants to hear. The factor of time in order to collect a representative sample of views, values, and opinions is also an issue. Questionnaires, although commonly utilized in attitude research, were also not chosen due to criticism they received in previous attitude research. Such criticism is based on the notion that attitude questionnaires attempt to reduce a multifaceted and interdependent construct to a few easily measureable quantitative dimensions. Because of this assumption, questionnaires typically reveal only the most evident attributes of attitudes and fail to render any underlying complexity of feelings or views.

Each of the three protocols consists of interview questions and provocative statements for individuals to respond to. These protocols provide the observation portion of the assessment triangle. The construction of these protocols included a preliminary phase during which trial questions and interview strategies were tested. Questions were derived from the principal themes being investigated—value of school science, how participants saw the value of applications of science to their everyday life, their visions of school science in the future, and the appeal and interest of science to them. Later these questions were modified based on responses. Provocative statements were added to the protocols in light of the trials revealing that forcing a reaction from the participants is more productive than discussing an experience (Osborne & Collins, 2000). Sample items from each of the three protocols are shown in Table 7.1. Because questions were derived from the main themes to be investigated, it seems clear that the interview protocols were developed with the purpose of assessment in mind.

Interpretation

Focus group interviews were recorded and transcribed for analysis. The transcriptions were then coded using a grounded theory approach for any emerging themes and issues (Osborne & Collins, 2001). This approach provides the interpretation component of the assessment triangle. A reliability check was conducted by independently coding the same transcript. Interviews were then transferred to a qualitative software package (NUD·IST) for further analysis of the codes. Since the intent of the instrument is to provide evidence that yields emerging themes of participants' attitudes, beliefs, and values, the grounded theory interpretation method employed appears to be appropriate for this assessment purpose.

TABLE 7.1 Sample Items From Each of The Three Focus Group Protocols

Protocol	Sample Items
Students	The science that I read in the newspaper and see on TV is of no interest to most people. (Statement)
	If I had a free hand to decide what young people learnt in school science, I would not change anything. (Statement)
Parents	Do you talk with your children about science related things, issues or events in your home? What issues have you discussed recently? (Question)
	In today's society science is one of the most important subjects to study at school. (Explain why/why not). (Statement)
Teachers	How do you respond to students' perceptions of the value of science in career terms rather than in cultural terms? (Question)
	School science does not equip students with adequate knowledge and skills necessary to understand and participate in discussion of scientific issues in society. (Statement)

Note: Adapted from "Pupils' and parents' views of the school science curriculum," by J. Osborne and S. Collins, 2000, London: King's College London.

Relationships Among Cognition, Observation, and Interpretation

An important question regarding Osborne and Collins's attitude assessment is, did the identified components of the assessment triangle (cognition, observation, and interpretation) link together to create a cohesive whole (see Figure 7.1)? As mentioned previously, an important critique of Osborne and Collins's focus group protocols is that these instruments appear to lack an explicit model of cognition and learning. By providing a clear, explicit model of cognition and learning, one can potentially obtain better clues about the types of tasks that are needed to elicit individuals' attitudes, beliefs, and views and the types of interpretation methods that are appropriate to transform responses about participants' views into meaningful assessment results. In turn, these results can provide a better means in which appropriate informed decisions can be made regarding revision of the curriculum under investigation. Thus, explicitly adopting a model of cognition would allow the developers to anticipate the types of knowledge and skills likely to be elicited by tasks with certain features and likely obtain relevant information about students' views (Pellegrino et al., 2001). On the other hand, the connections between observation and interpretation appear to be appropriate for the instrument's intended use. These connections can be further strengthened by applying an appropriate model of cognition and learning that provides a developmental perspective about how individuals develop competence within the assessed domain. Figure 7.2 provides a representative view of Osborne and Collins's attitude instrument in relation to the assessment triangle. The cognition compo-

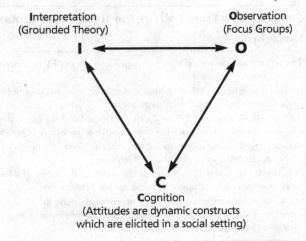

Figure 7.2 Osborne and Collins's (2000) attitude instrument (focus group interview protocols) in relation to the assessment triangle.

nent in Figure 7.2 has been purposely reduced in size to represent the lack of an explicit model of cognition and learning.

Although it is likely that the assessment developers, including Gogolin and Swartz (1992), did not consider applying the assessment triangle framework to their instrument design, we feel that by reflecting on their instrument with the assessment triangle as a lens will help improve the effectiveness of the attitude instrument.

Gogolin & Swartz (1992)

Context

Gogolin and Swartz (1992) investigated attitudes towards science of college student non-science majors along with students majoring in science. They made the distinction that they were focusing on attitudes towards science and not on scientific attitudes (Gardner, 1975). Specifically, the purpose of their investigation was to study "the attitudes toward science of nonscience college students to determine (a) how the attitudes toward science of nonscience students compare with the attitudes of science majors, (b) whether attitudes toward science change with instruction, and (c) what developmental experiences are associated with attitudes toward science" (Gogolin & Swartz, 1992, p. 488).

They cite several studies that indicate that science students have more positive attitudes towards science than non-science students in justifying their instrument and research design (Gabel, 1981; Korth, 1969; Shallis & Hills, 1975; Tilford, 1971). Therefore, they wanted to develop assessments to help them tease out the differences between these two groups of students. They investigated 102 college non-science students taking a terminal biology course in human physiology and 81 college science majors enrolled in the middle of a three-course biology sequence for majors.

Cognition

Gogolin and Swartz (1992) believed that the current attitude instruments available were limiting due to the results being distilled down to a single score. Therefore, they wanted assessments that could represent multiple influencing factors towards attitudes. Their study was framed around a model of learning that encompasses several educational environments including home, school and peer groups (Keeves, 1975), motivation, home environment and school environment (Kremer & Walberg, 1981), as well as the relationship between student, teacher, and learning environment (Haladyna, Olsen, & Shaughnessy, 1982). Their framework focused on interactions of these environments and the correlations they have with student learning. Their assessment instruments, described in detail in the next section, encompass these educational environments and interactions. A critique of the article, as analyzed through the COI framework, is that Gogolin and Swartz (1992) do not revisit their learning framework in the analysis. The attitudes toward science of both groups are reported, though only briefly, in the implications section providing there discussion of how these attitudes could affect achievement of these student groups. This could have been more explicit, and would have strengthened the cognition part of the assessment triangle making for a more complete assessment.

Observation

Gogolin and Swartz (1992) aimed for assessments that could provide a more complete view of students' attitudes towards science than those available. Therefore, they developed the Attitudes Toward Science Inventory (ATSI) as well as a questionnaire to uncover students' attitudes towards science along several subscales. The ATSI (modified from Sandman, 1973) is a multidimensional instrument including 48 four-point Likert-type items comprising six scales with eight items per scale. The six scales are (1) perceptions of the science teacher, (2) anxiety toward science, (3) value of science in society, (4) self-concept in science, (5) enjoyment of science, and (6) motivation in science. Representative sample items from the ATSI (Gogolin & Swartz, 1992, p. 491) are found in Table 7.2.

TABLE 7.2 Sample Items from the Attitudes Toward Science Inventory

Scales	Sample Items
Perceptions of the science teacher	Science teachers show little interest in the students.
	Science teachers know a lot about science.
	Science teachers don't like students to ask questions.
Anxiety toward Science	I feel at ease in a science class.
	It scares me to have to take a science class.
Value of science in society	Science is helpful in understanding today's world.
	Most of the ideas in science aren't very useful.
Self-concept in science	I usually understand what we are talking about in science class.
	No matter how hard I try, I cannot understand science.
Enjoyment of science	I would like to spend less time in school studying science.
	I enjoy talking to other people about science.
Motivation in science	I like the easy science assignments best.
	I would like to do some outside reading in science.
	The only reason I'm taking science is because I have to.

Note: Adapted from "A quantitative and qualitative inquiry into the attitudes toward science of nonscience college students," by L. Gogolin and F. Swartz, 1992, *Journal of Research in Science Teaching, 29,* p. 491.

In addition to the ATSI, a 45-question "interview questionnaire" (Gogolin & Swartz, 1992, p. 493) containing a mix of closed- and open-ended questions was developed and distributed to a random sample of 25 nonscience students (13 male, 12 female). Although throughout the analysis they refer to the interview data, it should be noted that interviews were not conducted. This assessment was clearly a written questionnaire. The questionnaire focused on four areas: (1) home environment, (2) school environment, (3) peer relationships, and 4) self-concept. Representative items from the questionnaire (Gogolin & Swartz, 1992) are found in Table 7.3.

The observation part of the assessment triangle was clearly defined and described in the article. Both assessments appeared to be linked to the learning model that framed their study, so the items in the ATSI do appear to represent the types of data that they were designed to collect. However, the questionnaire, if truly meant to be an interview, lacked true free re-

---

TABLE 7.3 Sample Items from the Interview Questionnaire

Interview Questionnaire

Family environment

　Did your family attend things such as science fairs or science museums when you were growing up?
　　☐ Quite often ☐ Occasionally ☐ Rarely ☐ Never

　Do your parents expect you to do good in school?
　　☐ Expect me to excel
　　☐ Do as well as I can
　　☐ Never say anything

　Can you think of anything in your home environment that has influenced how you feel about science?

School environment

　Did you do science experiments in elementary school?

　Were there science laboratories in your high school?

　Do you feel the science teachers you have had have been knowledgeable?

Peers

　Did your friends like science in elementary school?

　In high school, did your friends take biology, chemistry, math, and physics classes? List those classes taken

　Do you feel your friends had any influence on your choices of classes in high school?

Self

　How much control do you feel you have over what happens to you?
　　☐ Quite a lot ☐ Some ☐ Very Little ☐ None

　Would you do well in science if you tried to major in it?

　Why did you answer as you did?

Note: Adapted from "A quantitative and qualitative inquiry into the attitudes toward science of nonscience college students," by L. Gogolin and F. Swartz, 1992, *Journal of Research in Science Teaching, 29*, p. 493.

sponse questions as well as an opportunity to probe students with follow-up questions. Most of the items in the questionnaire could elicit yes/no responses, so the usefulness of this assessment is unclear.

Interpretation

The 48-item, 4-point Likert-type ATSI instrument was scored strongly agree (score = 4), agree, disagree, strongly disagree (score = 1) for each item. The scores were summed across the eight items for each scale, resulting in six attitude scores. Higher scores represent more positive attitudes, with the opposite considered for the anxiety dimension. Data from the ATSI were analyzed by multivariate and univariate statistics. For the questionnaire, the authors note that it was "used to probe for deeper understanding of these complex attitudes" (Gogolin & Swartz, 1992, p. 494). The results of their study revealed that there was a statistical difference between the non-science and science students' pre-test attitudes in all six scales. When compared with the post-test, they also found that both groups' attitudes changed significantly. The authors caution over-interpretation of the data and note, "although the mean scores were statistically different, the practical value of the differences remain to be determined. The amount of change that can be induced and measured in a 10-week period is unknown" (Gogolin & Swartz, 1992, p. 498). This distinction is important, because their results revealed that non-science attitudes increased (became more positive), whereas science student attitudes decreased (became less positive). Figure 7.3 provides a representative view of Gogolin and Swartz's

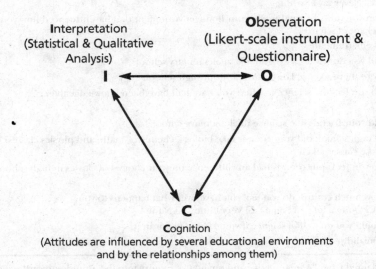

Figure 7.3 Gogolin and Swartz (1992) attitude instrument in relation to the assessment triangle.

(1992) attitude instrument in relation to the assessment triangle. In terms of strength and clarity, observation > interpretation > cognition. These components in Figure 7.3 have been purposely altered in size to represent this graphically.

In our analysis of the instruments, and what can be interpreted from them, it appears that the ATSI would be an effective instrument to gauge student attitudes towards science in a limited way, but not to compare groups. The instrument did not have multiple, equivalent versions to administer multiple times, so it is not ideal for investigating gains; however, this is not uncommon. The items were not well designed to distinguish between the science and non-science majors, as shown by their unanticipated results. The findings one obtains from the ATSI should refer back to the model of learning that framed the development of the instrument. This would strengthen the interpretation of the results. In addition, the questionnaire would need considerable revision to be a useful diagnostic instrument to elicit a deeper understanding of students' attitudes towards science. The questions should be reworded to remove the possibility of yes/no responses, and the data would certainly be richer if the instrument was actually used in real individual interviews.

A CLASSROOM APPLICATION OF ASSESSING ATTITUDES

Currently we are conducting a case study investigating students' attitudes toward science in relation to their epistemology and conceptual understanding of biotechnology. This study is an expansion of a pilot study we conducted earlier, examining students' epistemic beliefs and conceptual understanding (Rebello, Siegel, Freyermuth, & McClure, 2010; Rebello, Siegel, Witzig, Freyermuth, & McClure, under review). Participants in this study are undergraduate non-biochemistry majors enrolled in Biotechnology and Society, a large enrollment course.

By adopting Siegel and Ranney's (2003) perspective of attitudes toward science, we view attitudes as a construct that influences "cognition"—specifically, that "those who view science in a sophisticated way are better able than others to use their knowledge in more contexts and to make sense out of complex information" (Siegel & Ranney, 2003, p. 758). To explore students' epistemic beliefs, we utilized King and Kitchener's (2004) Reflective Judgment Model as our framework. We view students' epistemic beliefs based on how well they reason and use evidence in their decision making process. Students who demonstrate a higher epistemic belief are better able to apply knowledge they possess and reasoning to support their decisions (King & Kitchener, 2004). These two perspectives, attitude toward science and epistemic beliefs, together form the cognition element of our assessment triangle, and we are investigating relationships between these two perspectives that share similar groundings.

The observation element of our triangle includes the assessment of students' attitudes toward science and epistemic beliefs, using both quantitative and qualitative methodologies. First, to assess attitudes each of the case study participants were administered the Changes in Attitude about the Relevance of Science (CARS) questionnaire (Siegel & Ranney, 2003) electronically. The CARS questionnaire consists of 25 five-point Likert scale response items, including an "I don't understand" option. This questionnaire was specifically designed to measure students' attitudes toward the relevance of science to their daily lives as they make decisions (Seigel & Ranney, 2003). Second, to elicit students' epistemic beliefs we conducted individual interviews using King and Kitchener's (1994) reflective judgement protocol. This protocol was designed to elicit and assess epistemic beliefs according to their framework; specifically, the protocol focuses on how individuals view knowledge and justify decisions they make about controversial, ill-structured issues (King & Kitchener, 2004). Both the questionnaire and interview protocol were purposely designed to reflect the intended model of cognition and learning.

The interpretation element of our triangle includes interpreting the results from both the CARS questionnaire as well as the interviews. Interpretation of results from the CARS questionnaire is based on multiple analyses. Meanwhile, interpretation of the interviews is performed using inductive analysis. Each of the analysis procedures reflects the intended purpose of the instrument. That is to say, the interpretation method best reflects the cognition element being investigated and is appropriate for observations used in the assessment.

Reflecting upon the cognition, observation, and interpretation components of the assessment triangle, each well-defined component can have a significant influence on the design of an assessment instrument or an assessment study. Ensuring each component connects to the desired intentions of the assessment can provide an optimal means for obtaining meaningful results upon which informed decisions can be based.

DISCUSSION

Research on attitudes toward science often focuses on just one of the assessment triangle elements (cognition, observation, or interpretation). From our review of attitude assessments, much of the development and research of instruments focuses on types of observation methods. Little research has been conducted using various models of cognition and learning or various interpretation methods that can be usable by all.

For example, Osborne and Collins (2000) provide an interesting outlook on cognition, an area that typically has a limited perspective in attitudes research. They appear to view, although not explicitly, attitudes as a social, dynamic construct that cannot be measured in one simplified instance and has to be elicited in a social setting, such as focus groups. This is unlike another study conducted by Piburn and Baker (1993), who conducted individual semi-structured interviews to investigate factors that contribute to attitudes toward science, but had a different perspective for cognition. In this particular study, Piburn and Baker appear to view attitudes as mental constructs that students create from their own perspective. Viewing these constructs through teacher report or classroom observation is seen as inadequate (Piburn & Baker, 1993). Based on our reviews, researchers generally take a similar individualized perspective of attitudes. However, unlike Osborne and Collins, little research in attitudes toward science had undergone a socio-cultural perspective. Today, research in science education and other domains has shifted more toward research in team cognition and learning communities. Yet, research on attitudes toward science still focuses on the individual. If attitude research is to become more in line with these shifting cognition perspectives, more research with these perspectives (e.g. socio-culture perspectives) has to be conducted. Also, these models of cognition and learning need to be explicitly stated in the assessment design to help map the development of connections of remaining assessment components.

Another limiting aspect of attitude research is the variety of instruments (observations) employed to assess attitudes. Because there tends to be a general cognition outlook in the area of attitudes, most of the instruments available are only limited to that particular perspective. Many of the instruments available are conducive for research in attitudes; however, the question remains: Are these instruments employable in a classroom setting by instructors? Instructors may want different varieties of assessments that are easily employable in their classrooms, yet little research has discussed this topic. Such alternative assessments could be: students interviewing each other with an appropriate protocol or an open-ended assessment that allows students to reflect on their learning.

Interpretation methods are another factor in attitudes research. Research generally focuses on advanced measurement techniques for interpretation of results. Yet there is a gap between the researchers' interpretation methods and that of the instructors. Most of these interpretation methods are not doable by instructors and provide little meaning for instructors. Therefore, there is a need to provide an interpretation method that is effective yet quick and easy for instructors to perform in a classroom setting, thus allowing instructors to obtain meaningful feedback in a timely manner for educational purposes.

Not only is the focus of each element important, but the connections between the cognition, observation, and interpretation are important as well. Without these connections, an assessment may not provide meaningful evidence in which appropriate inferences can be made (Pellegrino et al., 2001). This may impact important decisions made at all levels of education, especially in the classroom or at the policy level. A lack of connection between the assessment triangle elements appears to be evident in the case of Gogolin and Swartz (1992). In this particular case, it appears that the authors did not make a clear connection between their learning framework (cognition) and their analysis of results (interpretation). If this connection were explicitly made, the learning framework could have the potential to provide better insights toward what interpretation methods are appropriate to use to draw meaningful inferences from the data.

Research on attitudes toward science often focuses on just one of the assessment triangle elements, rather than all three. A focus on one element can be important in order to make progress in a research area, but more studies are needed that attend to each element and the connection between them. In this book, a few examples of chapters that focus on one element include: Al-Salih, who emphasizes the cognition element in a discussion of theories; Hren, who focuses on the observation element with discussion of a new instrument; and Yu and Moore Menash, who emphasize the interpretation element through the multiple response model. Psychometrics research and the assessment triangle are not different from each other, especially concerning the need of establishing validity. The assessment triangle simply provides a guide to designing a valid instrument that supports reasoning from evidence. Specifically, the evidence and theory support the interpretations made from the assessment results.

REFERENCES

Bennett, J. (2001). The development and use of an instrument to assess students' attitude to the study of chemistry. *International Journal of Science Education, 23*(8), 833–845.

Blalock, C. L., Lichtenstein, M. J., Owen, S., Pruski, L., Marshall, C., & Toepperwein, M. (2008). In pursuit of validity: A comprehensive review of science attitude instruments 1935–2005. *International Journal of Science Education, 30*(7), 961–977.

Gabel, R. K. (1981). Attitudes toward science and science teaching of undergraduates according to major and number of science courses taken and the effect of two courses. *School Science and Mathematics, 1*, 70–76.

Gardner, P. L. (1975). Attitudes to science: A review. *Studies in Science Education, 2*, 1–41.

George, R. (2000). Measuring change in students' attitudes toward science over time: An application of latent variable growth modeling. *Journal of Science Education and Technology, 9*(3), 213–225.

Gogolin, L., & Swartz, F. (1992). A quantitative and qualitative inquiry into the attitudes toward science of nonscience college students. *Journal of Research in Science Teaching, 29*(5), 487–504.

Haladyna, T., Olsen, & Shaughnessy, J. (1982). Relations of student, teacher, and learning environment variables to attitudes toward science. *Science Education, 66,* 671–687.

Keeves, J. P. (1975). The home, the school, and achievement in mathematics and science. *Science Education, 59,* 439–460.

Kind, P., Jones, K., & Barmby, P. (2007). Developing attitudes towards science measures. *International Journal of Science Education, 29*(7), 871–893.

King, P. M., & Kitchener, K. S. (Eds.). (1994). *Developing reflective judgment: Understanding and promoting intellectual growth and critical thinking in adolescents and adults.* San Francisco: Jossey-Bass.

King, P., & Kitchener, K. S. (2004). Reflective judgment: Theory and research on the development of epistemic assumptions through adulthood. *Educational Psychologist, 39,* 5–18.

Koballa, T. & Glynn, S. (2007). Attitudinal and motivational constructs in science learning. In S. Abell & N. Lederman (Eds.), *Handbook of research on science education* (pp. 75–102). Mahwah, NJ: Lawrence Erlbaum Associates.

Korth, W. W. (1969). *Test every senior project: Understanding the social aspects of science.* Bethesda, MD: U. S. Department of Health, Education, and Welfare. (ERIC Document Reproduction Service No. ED 028 087)

Kremer, B. K., & Walberg, H. J. (1981). A synthesis of social and psychological influences on learning. *Science Education, 65,* 11–23.

Munby, H. (1997). Issues of validity in science attitude measurement. *Journal of Research in Science Teaching, 34*(4), 337–341.

Osborne, J. F. & Collins, S. (2000). *Pupils' and parents' views of the school science curriculum.* London: King's College London.

Osborne, J. & Collins, S. (2001). Pupils' views of the role and value of the science curriculum: A focus-group study. *International Journal of Science Education, 23*(5), 441–467.

Osborne, J., Simon, S., & Collins, S. (2003). Attitudes towards science: A review of the literature and its implications. *International Journal of Science Education, 25*(9), 1049.

Pellegrino, J. W., Chudowsky, N., & Glaser, R. (Eds.). (2001). *Knowing what students know: The science and design of educational assessment.* Washington, DC: National Research Council.

Piburn, M. D. & Baker, D. R. (1993). If I were the teacher: Qualitative study of attitude towards science. *Science Education, 77*(4), 393–406.

Ramsden, J. M. (1998). Mission impossible? Can anything be done about attitudes to science? *International Journal of Science Education, 20,* 125–137.

Rebello C. M., Siegel M. A., Freyermuth S. K., & McClure B. A. (2010). "Genetically modified foods are the only foods that have DNA": Epistemological beliefs and conceptual understanding in a non-majors biotechnology course. *Pro-*

ceedings of the Annual Meeting of the National Association for Research in Science Teaching, Philadelphia, PA.

Rebello C. M., Siegel M. A., Witzig S. B., Freyermuth S. K., & McClure B. (in press). Epistemic beliefs and conceptual understanding in biotechnology: A case study. Submitted to the *Research in Science Education.*

Reid, N. (2006). Thoughts on attitude measurement. *Research in Science & Technological Education, 24*(1), 3–27.

Sandman, R. S. (1973). *The development, validation, and application of a multidimensional mathematics attitude instrument.* Unpublished doctoral dissertation, University of Minnesota.

Schibeci, R. A. (1984). Attitudes to science: An update. *Studies in Science Education, 11,* 26–59.

Shallis, M. & Hills, P. (1975). Young people's image of the scientist. *Impact of Science on Society, 25,* 275–278.

Shrigley, R. L., Koballa, T. R., & Simpson, R. (1988). Defining attitude for science educators. *Journal of Research in Science Teaching,* 25, 659–678.

Siegel, M. A., & Ranney, M. A. (2003). Developing the changes in attitude about the relevance of science (CARS) questionnaire and assessing two high school science classes. *Journal of Research in Science Teaching, 40*(8), 757–775.

Tilford, M. P. (1971). Factors related to the choice of science as a major among negro college students. Doctoral dissertation, Oklahoma State University. *Dissertation Abstracts International, 33,* 5970A.

Tomanek, D., Talanquer, V., & Novodvorsky, I. (2008). What do science teachers consider when selecting formative assessment tasks? *Journal of Research in Science Teaching, 45*(10), 1113–1130.

CHAPTER 8

THE INFLUENCE OF EXPERIENTIAL LEARNING ON INDIGENOUS NEW ZEALANDERS' ATTITUDE TOWARDS SCIENCE

Enculturation into Science by Means of Legitimate Peripheral Participation

Richard K. Coll and Levinia Paku
University of Waikato, New Zealand

ABSTRACT

Worldwide indigenous peoples are reported to be underrepresented in science and engineering higher education study and related careers. The literature suggests this is due in part at least to indigenous people feeling alienated from Western ideas of science, with entry into science viewed as a form of "border-crossing" into a new cultural community. Recent research in our group suggests indigenous New Zealanders who engage in experiential learning as part

Attitude Research in Science Education, pages 219–238

of a work-integrated learning program in science and engineering, like their non-indigenous counterparts, are rapidly enculturated into the community of practice that forms the scientific community. This, it appears, occurs by means of *legitimate peripheral participation* in the community as these "newcomers" work alongside practicing scientists in a form of cognitive apprenticeship. In doing so they gradually adopt features of the scientific attitude and enhance their attitude towards science, eventually seeing themselves as legitimate members of the scientific community and developing a sense of belonging that is not easily achieved in conventional programs of study in higher education.

INTRODUCTION

Worldwide, there has been much concern expressed about falling student interest in science, science enrollments at higher education providers, and lack of interest in or uptake of science careers (Coll, 1996; Coll & Eames, 2008; Fensham, 1980). A number of reasons for such problems with science have been proposed. Science is seen as an unpopular or difficult topic for school study at the primary and secondary level, and science careers are seen as less financially rewarding or glamorous than careers in medicine, law, or business studies (Dalgety & Coll, 2004, 2005). Thus, it seems science has something of an image problem, with scientists perceived as engaging in dangerous or unpopular work (e.g., biotechnology, Brunton & Coll, 2005), odd or unusual, or socially inept (Dalgety & Coll, 2004), and much of the general public generally adheres to very stereotypical images of scientists (see, e.g., Coll & Taylor, 2004; Dalgety & Coll, 2004). Much of this negative view of science seems image-driven, and a constant barrage of negative images of science and scientists likely influences students' perceptions of, and attitude toward, science and scientists. While the literature suggests that many attitudinal views are generated at primary school (see Coll & Taylor, 2004), interestingly, according to Dalgety and Coll (2005), this is the case even for students engaged in higher education. In this work, we argue that science is seen as alien to many students, and scientists are seen as belonging to an exclusive club or cultural group. Here we explore the nature of this science culture and describe how students working alongside professional scientists as part of purpose-designed, work-integrated learning programs help students, including students from indigenous cultures, feel a sense of belonging to the culture of science.

LEARNING SCIENCE

There are numerous studies reported in the science education literature that suggest science is a subject that many students find difficult to learn.

Duit and co-workers have compiled extensive bibliographies of investigations into student alternative conceptions for science conceptions (i.e., student conceptions that agree in disagreement with consensually agreed scientific views) (Duit, 2007, 2009; Pfundt & Duit, 1994, 1996, 2000). Much of the alternative conceptions research has sought to understand why student alternative conceptions are so prevalent, and the origins of such alternative conceptions. As might be expected, numerous theories have been posited. Much early research focused on the learning process itself, as the alternative conceptions movement is strongly associated with the development of constructivism and related theories of learning. Constructivists say that traditional ways of science teaching assumed the learning process was non-problematic, and that any student failure to learn was due to laziness, inattention, or lack of cognitive development (Tobin & Tippins, 1993). However, constructivism suggests that learning is instead a process of students constructing new knowledge in their own minds, that this mental construction is strongly influenced by the student's prior knowledge, and that it is socially mediated (Good, Wandersee, & St. Julien, 1993). In other words, science learning is situated in a particular social context (i.e., a school, university, etc.), and what is "good" or true knowledge is that which is consensually agreed by members of that community (Lave, 1991; Lave & Wenger, 1991).

SCIENCE AS CULTURE

Science and scientific thinking is so different from everyday life that it has been described as being a culture in and of itself (Aikenhead, 2001). If this is, in fact, the case, then for someone to become a scientist, one must learn about and become part of that culture, or cultural community—what Lave (1991) describes as a *community of practice*. According to Aikenhead, this is deeply problematic for many students, and it involves what he terms "border crossing," in which students are thought to cross cultural "borders" from their life world subculture into the subculture of science. This transition or border crossing is difficult for any student, even when school science beliefs are aligned with Western beliefs and they come from a Western cultural background (Cobern & Aikenhead, 1998). However, for indigenous peoples, such a transition is likely to be much more problematic (Glasson, Mhango, Phiri, & Lainer, 2010; Keane, 2008; Paku, Coll, & Zegwaard, 2003). Traditional research about enculturation involves notions of assimilation (Aikenhead, 1998), according to which enculturation occurs when science instruction has been embraced and it is in agreement with students' view of the world—meaning the student becomes embedded in the science that is being taught. However, if the culture of science is seen to be very different from the culture of the learner (as it is suggested is the case for indigenous persons—Glasson et al., 2010; Paku et al., 2003), then this results in *assimilation* where Western science is *inflicted* on indigenous students, and all other beliefs of science must be abandoned in the process. Assimilation therefore raises issues about equity and emancipation. The literature thus suggests that the notion of border crossing or enculturation is likely to be more problematic for indigenous peoples studying science or considering careers in science.

Cooperative education or work-integrated learning (WIL), we suggest, allows students to develop an understanding of the nature of science during practice, and if this is done in a positive way, such learning may exert a positive influence on their attitude toward science. It seems unlikely that students, indigenous or otherwise, will develop positive feelings about science and scientists if they feel alienated from this subculture. If one accepts that to understand science one needs to actually *do* some science, then programs of study that allow students to do science and work alongside professional scientists have considerable scope to enhance student understanding of the nature of science, and this may smooth their enculturation into science (Eames & Bell, 2005). Moreover, WIL allows students to be involved in decision making, problem solving, and being able to make informed choices about the science they are doing. This not only gives the student experience within the science subculture, but it gives them an understanding of what it means to be a scientist.

STUDENT LEARNING EXPERIENCES AND STUDENT ATTITUDE TOWARDS SCIENCE

Student attitude towards science has been extensively studied and reviewed (Bennett, Rollnick, Green & White, 2001; Gardner, 1975; Koballa, 1990; Schibeci, 1984). As noted above, it seems students think of scientists in terms of the "mad scientist" stereotype (e.g., Billingsley, 2000; Jones, Howe & Rua, 2000; Schibeci, 1986), but some students, particularly females, think scientists are less anti-social than the stereotype presumes (Lips, 1992). While some students identify scientists as people who help society, particularly in relation to environmental issues and medicine (Dawson & O'Connor, 1991; Koballa, 1990), science is perceived by others to be destructive to civilization and to contribute to the creation of societal problems (Jones et al., 2000). There also is some confusion regarding the nature of science and the nature of scientific research (Allchin, 1990; Krasilchik, 1990; Lederman, 1992). For example, students do not view science as a social activity—knowledge validation and choices of research are seen to be preordained and immutable, rather than socially negotiated via peer-review, or as beliefs held by a community of practitioners (Ryder, Leach & Driver, 1999).

Research on the influence of context on attitude towards science has focused on gender, ethnicity, and the effect of age and intelligence (Koballa, 1990). It seems there is little difference in the strength and polarity of male and female attitudes towards science (Piburn, 1993; Rennie & Dunne, 1994; Thompson & Soyibo, 2002), suggesting the widespread gender differences Gardner (1975) reported in an early review of attitude towards science are nowadays less prevalent. More recent research suggests that other factors that influence student attitude towards science include age and exposure to science (Myers & Fouts, 1992). For example, rather worryingly, as students get older and have increasingly more choice in participating in science activities, their attitude towards science becomes less positive (Butler, 1999; Gogolin & Swartz, 1992).

Students' learning experiences and experiences with teachers influence their attitude towards science at all levels of study (Speering & Rennie, 1996; Talton & Simpson, 1987). Teacher–student interaction, as part of student learning experiences, influences student attitude towards science (Schibeci & Riley, 1986). In particular, students with negative attitude towards science are influenced by negative perceptions of their teachers (Gogolin & Swartz, 1992). Teacher attributes that result in negative attitudes in the transition between middle and secondary school include less active teaching style, strategies, and teaching approaches (Gibson & Chase, 2002; Speering & Rennie, 1996). More active learning approaches such as project work have been found to increase students' understanding of how scientists conduct research (Ryder & Leach, 1999). For example, higher-level undergraduate university students were more aware of the length of time it took to carry out research and the systematic approach of data collection after working on science projects. The highly contextualized project work increases awareness of the importance of socialization and group work in science (Jarvis & Pell, 2001, 2002a, 2002b).

Science is often perceived as difficult compared with other subjects such as in the humanities. This may be due, in part, to the belief that students must obtain the "correct" answer—during summative assessment at least (Dawson & O'Connor, 1991). Yet, studies of student science self-efficacy (perception of their ability to undertake science tasks) are sparse, although there has been some research practice in this area (see, e.g., Yang, Andre & Whigham, 2001). There are a number of factors that influence self-efficacy, including contextual variables such as school learning environments (Lorsbach & Jinks, 1999; Tymms, 1997), and it is reported that science self-efficacy is related to individual school variables. Science self-efficacy is also influenced by student attitude towards science (Jones & Young, 1995; Smist & Owen, 1994; Talton & Simpson, 1986). If students perceive that scientists are "normal," this positively influences their self-efficacy, meaning they are more confident about engaging in science learning (Smist & Owen, 1994).

Vicarious experiences have been reported to be effective in improving students' self-efficacy (Luzzo, Hasper, Albert, Bibby, & Martinelli, 1999). Interventions reported to be effective include tests designed specifically to be easy to pass—intended to build students' confidence about their ability and thus increase their self-efficacy for a given topic or task (Chang & Bell, 2002). This, it seems, is more effective other than vicarious experiences, such as exposure to "successful" students. This latter strategy seeks to enhance student self-efficacy by enabling them to compare themselves with others—of comparable or lesser ability than themselves—who have proven to be successful (Anderman & Young, 1994; Luzzo et al., 1999).

In summary, research into students' learning of science suggests that learning science and becoming a scientist involves becoming a member of the science subculture. Their border crossing is influenced by their attitude towards science, which is in turn influenced by their self-efficacy and science learning experiences. Research about learning suggests that student enjoy more active learning approaches. Next we describe WIL in more detail—a learning approach that involves students working with scientists and learning about the culture of science where they learn by doing science.

WORK-INTEGRATED LEARNING PROGRAMS

Work-integrated learning (WIL) is a collaborative enterprise in which students, employers, and education providers work together to produce work-ready graduates—that is, graduates who have (particularly) practical skills that complement theoretical academic learning and that make them of almost immediate value to employers (Groenewald, 2004). WIL has a variety of names, and these often reflect its location. So in the UK it is seen most commonly in "sandwich" programs, and in the U.S., it takes the form of cooperative education (co-op) work placements spaced evenly throughout the degree, or capstone internships—placements added on to the end of a degree (e.g., in medicine). In any system, the students spend predetermined periods of time—such periods of time commonly called work placements—in a relevant workplace. So an engineering student at university might complete two three-month work placements in an engineering firm, a food technology student at a polytechnic or community college might do the placement in the form of one day a week in a food testing laboratory, and so on.

The origins of WIL as an educational strategy lie in the U.S., in particular in the University of Cincinnati (UoC). Herman Schneider of the UoC is generally regarded as the founder of the system (although it is acknowledged that the idea of combining work and study has earlier exemplifications—

see Walters, 1947), probably because he coined the term and championed the notion of work-integrated learning at the institutional level—something highly novel at the time. Schneider termed this cooperative education, but nowadays, the broader term used to describe this learning system is *work-integrated learning* (WIL) (Franks & Blomqvist, 2004). In any case, the identifying feature of WIL is a combination of work experience with on-campus academic learning, but also the notion that learning from each site is *integrated*. So, for example, a student studying analytical chemistry at a university might take knowledge learned from his or her studies and use this to engage in meaningful work with an employer who runs an analytical service laboratory.[1] But at the same time, the student might take perhaps current research ideas based on leading edge research in a particular analytical technique (e.g., modern instrumental techniques) learned from university study into the work force. Even if students do not take such knowledge into industry, they take a different culture or way of thinking, and this can form part of the integration in work-integrated learning (Coll, 2004). The extent to which any integration actually occurs has been the subject of a recent major, national, cross-industry study (see Coll et al., 2008), which indicated that integration is much more ad hoc in nature than we might hope.

Commitment to WIL by employers, education, and governments waxes and wanes somewhat, although it is currently very much in vogue. After a slowish start, WIL underwent massive expansion in the U.S. as a result of substantial commitment from federal authorities (Ryder & Wilson, 1987), but this abated dramatically when the rapid growth resulted in program quality issues, as many educational institutions strived to gain their share of this largesse (Sovilla & Varty, 2004). Outside the U.S., growth has been less spectacular, but probably more measured in nature, and in recent times WIL has expanded into Asia in particular (see, e.g., Coll, Pinyonatthargarn & Pramoolsook, 2003, 2004). The benefits of WIL across all three sectors are now well recognized, so that many governments see it as a means of preparation of work-ready graduates—as exemplified by Dearing's recommendation that all UK degrees should incorporate some work experience (Dearing, 1997).

WIL in one form or another is thus now a major "industry" worldwide and is practiced widely in the U.S., UK, Australia, the Asia-Pacific region, South Africa, and Europe (Franks & Blomqvist, 2004). It is interesting to consider why WIL has been so successful, and yet maintains a relatively low public profile. WIL is something of an intuitive concept, one that appeals to almost anyone engaged in hiring employees or training graduates (Eames, 2003). The principal argument is that all three parties involved in cooperative education (i.e., students, employers, and education providers) stand to benefit (Franks & Blomqvist, 2004). It also now has a fairly substantial research base that supports anecdotally claimed operational outcomes (Bartkus & Stull, 2004).[2] Dressler and Keeling (2004) provide a substantial list of research studies that report students accruing a variety of benefits as a consequence of WIL over conventional programs. These consist of *academic benefits* (e.g., increased motivation to learn, increased ability to finance tuition, improved perception of benefits of study, etc.), along with *personal benefits* (e.g., increased autonomy, increased communication skills, improved time management, etc.) and *career benefits* (e.g., increased employment opportunities, career clarification, international opportunities, etc.). Employer benefits, as might be expected, are fairly pragmatic in nature and mostly concern the work-readiness of graduates (Braunstein & Loken, 2004). Overall employer benefits are financial in nature (lower recruitment costs, increased productivity, etc.) but also issues to do with image (e.g., addressing equity in employment, enhanced public image of major multi-national corporates, etc.). Reported benefits for education providers again are fairly pragmatic, and include things such as enhanced student recruitment, stakeholder input into program development, and enhanced links to industry, with the latter often resulting in ongoing, commercially beneficial relationships (Weisz & Chapman, 2004).

WORK-INTEGRATED LEARNING AS AN EDUCATIONAL STRATEGY

The literature thus suggests WIL is beneficial, in mainly operational terms, for all three parties. However, as noted above, the practice of what comes under the umbrella term "work-integrated learning" varies substantially, and this has consequences for WIL as an educational strategy. A key issue here is the purpose of the particular WIL program. This, not surprisingly, is related to perceived benefits mentioned above. So the purpose of WIL from an employer's viewpoint is largely vocational (Dressler & Keeling, 2004); hence, the educational purpose here is *vocational training*. A similar view is held by many education providers, who see a similar purpose, and the fact that WIL graduates gain work more easily fits in nicely with their recruitment strategies. However, recently Coll and Eames (2004) have argued that WIL needs to be more broadly educational in nature. That is to say, it "is about learning" (p. 273), even if the *outcomes* are pragmatic and vocationally oriented. This, it is argued, allows WIL to function as a much broader, more holistic educational strategy:

> Only by employing strong curricula and pedagogy underpinned by theory, and objectives that are *relevant and appropriate to all parties* involved can a successful co-op [or WIL] program be sustained. We are not here advocating total focus on the student (although the role of co-op [WIL] placement coordinators might well drive them to such a position); we feel that a balance

is necessary, and have some concern that a program that is too vocationally focused might well lose sight of the student. Students *do* need to be equipped with skills that will help them find meaningful employment (something which will likely satisfy employers), but surely education is broader than this. . . . Our point is that in our view we do need to maintain clear *educational* goals in co-op [WIL]. The objectives set for the co-op [WIL] program and the work component should allow the student to engage in critical thinking and transformative learning. We recognize the tension that exists between the primary goals of academia (education) and industry (productivity), but we feel education must remain paramount. (Coll & Eames, 2004, pp. 274–275, original emphasis)

Eames elsewhere argues that WIL has not had a strong focus on learning, and that little research is reported of *what* and *how* students learn (see Eames, 2003; Eames & Bell, 2005). This is crucial if we are to run sustainable, mutually beneficial WIL programs, as Coll and Eames (2004) argue: "We need to know a lot more about what happens in terms of the 'what' and 'how' of co-op [WIL] *learning* that occurs before we can begin to design curricula and decide appropriate pedagogies" (p. 277). Eames (2003) argues that for WIL to be sustainable, it needs to be seen and *respected* as a key "value-added" educational strategy, at the institutional and governmental levels, and not just as a useful way of producing work-ready graduates. This has proven problematic, with the administration of WIL often scattered throughout a given institution, and with few institutions having a coordinated, coherent, institution-wide approach to WIL. It is worthwhile, however, to note that there are several North American-based education providers that have fully embraced WIL as an educational strategy, namely Northeastern University (Northeastern University, 2007), and Waterloo University in Canada (McLaughlin, 1997), both of who offer WIL at the institutional level.

RESEARCH INTO EDUCATIONAL ASPECTS OF WORK-INTEGRATED LEARNING PROGRAMS

Research in WIL took a step forward with the publication of the first major study of the learning that occurs in WIL programs by Eames. As noted above, WIL research has been criticized for a lack of theoretical base (see Bartkus & Stull, 2004; Eames & Cates, 2004); here we the mean theoretical basis to *learning*. Eames's work helped shift the research agenda to a new, more rigorous, level for several reasons. First, it provided the "missing link"—namely, a solid theoretical base to WIL research—drawing on sociocultural theories of learning (Eames & Bell, 2005). Eames argues that to understand WIL as an educational strategy, we need to be cognizant of the importance of contextual factors, especially sociological factors. So he talks of the student learning to become a scientist via *legitimate peripheral participation* (Lave & Wenger, 1991), as he or she works alongside scientific experts. This opportunity to appropriate the knowledge, skills, and culture of that scientific workplace leads to a deeper understanding of what it means to work in science (Rogoff, 1995). He then speaks of the notion of *mediated action*, in which learning in the workplace is a feature of the particular social circumstances. For example, language acts as a Vygotskyian tool (Vygotsky, 1978), meaning that there is a way of using language (e.g., writing or speaking "scientifically," such as the use of acronyms) that is specific to the socio-cultural context in which learning occurs. So acronyms like LCMS, GLC, and NMR are normal "language" in a chemistry laboratory, whereas terms like ATP and ADP are common in a biology laboratory. He also notes the concept of *distributed cognition* (Perkins, 1997) where knowledge is not resident solely in an individual (e.g., the workplace supervisor), but is distributed across the workplace. So, for example, a scientist in the organization might hold knowledge about how to conduct scientific research, but the technicians hold knowledge of specific instrument operation and maintenance, administrators about management and workplace OSH policies, and so on. A work placement thus allows a science student to develop an identity and become *enculturated* within a science community of practice (Lave & Wenger, 1991).

A particular feature of Eames's study is its longitudinal nature (the work spanned some four to five years, tracing a cohort of students all the way through their undergraduate WIL degree). The study also provides a fascinating insight into what and how learning occurs in a variety of scientific enterprises. Among the insights gained were that students gain an in-depth understanding of the research process (e.g., learning that research agenda shift depending on circumstances; the role of financing in research communities), enabling them to learn the behavior of a researcher, and that they develop a way of working and thinking in accordance with a research culture. Eames's students also were reported to have developed a deep sense of how they learned, and from whom, they learned (i.e., a developing sense of metacognition). They expected to learn from scientists, but were surprised to learn many useful things from others in the learning community in which they were situated (e.g., the technicians and office staff mentioned above). Arguably the most important understanding to come from this work was how different the socio-cultural environment was in the workplace compared with the academic learning environment. Two examples illustrate some important differences. First, in the workplace, one-on-one interactions were routine; in the education provider this was rare, except in the advanced stages of the degree program. Second, in the workplace, particularly in the case of students placed into research institutes, students encountered learning experiences in terms of practical science that were totally different. In the university setting, the laboratory classes were highly organized, the way to conduct an experiment was highly detailed in the course laboratory manual, and there was an expected (often numerical) outcome, known in advance. This is in almost dialectic contrast with genuine research in a research institute, where the answer was most certainly not known—or even necessarily attainable, and the way to get "the answer" needed to be developed from scratch!

Evidence that Eames's work has indeed helped shift the research agenda in WIL research comes from a number of recent studies that have drawn upon his work.[3] This is quite varied and shows the utilitarian nature of the theoretical framework that he developed for WIL inquiry. The research mostly tries to understand science learning from a student viewpoint and to develop an understanding of their learning experiences that influence their attitude towards science and scientists. Arguably, then, it is not about learning *per se*, but about how people think they learn, what they see as barriers and enabling factors in their science learning, and about how they enjoy science learning, and so on. Two examples illustrate how this research has developed. The first looks at practical learning experiences; the second is developed in more depth in a subsequent section, and looks at the nature of enculturation into the science community of practice.

The capacity of education providers to provide real, in-depth learning experiences in practical science is debated extensively in the literature. In a substantial review, Nakhleh, Olles, and Malina (2002) come to the conclusion that there are very mixed results when it comes to justification for and convincing evidence that practical science experiences in school or higher education laboratories actually provide measurable learning outcomes. Other research suggests that this might be due to students' apprehension about engaging in practical science. Fletcher (1990, 1991) suggests WIL can enhance self-efficacy (a students' perception of their own capability about engaging in a specific task) and encourage learning by a process of enactive mastery (i.e., as students with sound mentoring are scaffolded though their learning in the workplace they gain in confidence as they "master" tasks). These positive experiences will then enhance student attitudes toward science. Coll, Lay, and Zegwaard (2001) subsequently looked at the influence of WIL on student perceptions of their ability in practical science. This, it is argued, enhances student self-efficacy, and thereby practical science skills in a number of ways, all of which influences their attitude toward science. Students gain relatively few practical skills at their education provider. So, as might be expected, they are typically very nervous about using expensive scientific instruments when they first start in industry placements. However, as they practice, under good supervision, they gain in skill and in self-efficacy. This is mediated by good mentoring enabled by *verbal persuasion*

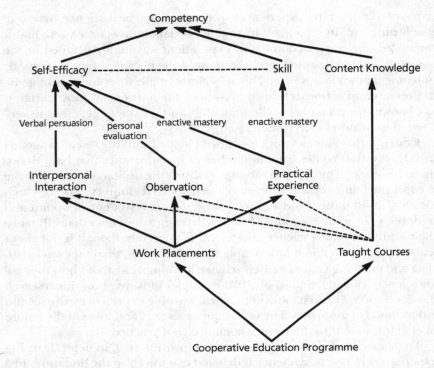

Figure 8.1 Relationship between self-efficacy, learning and cooperative education/work-integrated learning (after Coll et al., 2001).

(i.e., positive encouragement from their mentors), and *personal valuation* of their own capabilities—see Figure 8.1). If one accepts that enhancing self-efficacy is an enabling feature of learning, then this provides an example of the educational value WIL work experience can provide in science learning, achieving competency in practical science and in enhancing their attitude toward science. It is interesting how complex this process appears in the dual-situated learning environment that makes up the WIL experience as presented in the proposed theoretical framework.

WORK-INTEGRATED LEARNING PROGRAMS AND ENCULTURATION INTO SCIENCE

There has been much research about underrepresentation of certain student cohorts, such as women (Hanson, 1996) and indigenous peoples (Aikenhead, 2001), in science and technology study in higher education, and in science and technology careers (Cobern, 1998; Hanson, 1996). Paku,

Zegwaard, and Coll (2003) looked at the role WIL might play in enhancing access to science and technology careers, drawing on socio-cultural notions of learning proposed by Eames. As noted above, previous research into this area has been concerned with issues such as "border crossing" (Aikenhead, 1996) in which students are thought to cross cultural borders from their life-world subculture to the subculture of science and science learning.

Traditional research of "enculturation" (where a student's life-world subcultures harmonize with the subculture of science) and "assimilation" (of students whose life-world cultures differ from the subculture of science) also have been addressed in the literature (Aikenhead, 1996). At the institution in which Paku and colleagues (2003) worked, the approach to improve representation centered on the development of culturally appropriate support structures for Māori (i.e., individuals who self-identify as indigenous New Zealanders). Here the support structure comprises of *kaitiaki* (mentors). These *kaitiaki* are graduate students and are allocated undergraduate students with similar degree interests. The support given is focused on guidance, providing students with direction and helping them to become "university wise." Contact is maintained by encouraging *kanohi ki te kanohi* approach (face-to-face approach) throughout laboratory sessions and regular *hui* (meetings) throughout the year. Students are also encouraged to take advantage of a resource room that is available to them, which is also an opportunity to meet with other Māori and utilize the facilities available to them there. All of these initiatives are driven by a consideration of Māori culture, values, ways of thinking—their emphasis on a sense of belonging, and the impact this has on learning. So how can WIL fit into, or enhance, such support structure? Paku and colleagues (2003) report that it is clear that Māori students that completed a WIL placement had gained considerable understanding during this period, strongly influenced by the type of knowledge, background of study and type of placement. They freely used terms common to their scientific discipline and those gained from the work placement to describe broader terms such as science and technology, and this influenced how they viewed the roles of technologist and scientist.[4] Students who had completed work placements found the experience particularly useful, feeling that their understanding and enthusiasm for their study of discipline had improved, and that the understanding of the relevance of aspects of their studies complemented the practical component of their discipline. Perhaps most importantly, these Māori students were far more positive about science learning and embarking on science careers; they now felt there was a place for them in science and in science careers, the net result being an enhanced sense of belonging.

SUMMARY AND CONCLUSIONS

Student attitudes toward science and science careers is a key issue for many nations. Science and technology are seen as key enablers of economic development, and there is real concern about falling interest in further science study and science-related careers. Extensive research suggests that many students find learning science problematic, mostly as a result of highly teacher-centered, passive learning experiences. More active learning experiences have the possibility of improving student perceptions of science self-efficacy, enjoyment of science, and ultimately their attitude toward science. To learn science or to become a scientist involves learning about the subculture of science and becoming a member of the science community or subculture. This crossing over into the science subculture is particularly problematic for indigenous peoples, whose own culture is seen as incompatible or in conflict with, the Western subculture of science. Work-integrated learning programs, in which students work alongside professional scientists by means of legitimate peripheral participation, can ease indigenous and non-indigenous students' enculturation into the subculture of science. This occurs by means of verbal persuasion from supportive mentors, and personal evaluation of their own improving practical science skills as they engage in work placements.

NOTES

1. This need not necessarily be *paid* work, but it needs to be of value to the employer (see Coll, Eames, Zegwaard & Hodges, 2002).
2. Operational outcomes are practical things like, do employers find that students who graduate from cooperative education come up to speed more quickly? Do graduates earn more, or advance more rapidly in their careers?, etc.
3. Eames's work also has contributed to a more theoretically-rigorous research agenda in co-op research internationally (see, e.g., Eames, 2000), and in particular in the from of an invited paper as a lead paper in the 2006 issue of the *Journal of Cooperative Education & Internships*—see Eames and Coll (2006).
4. Interestingly, this fits in closely with Eames's (2003) findings for non-Māori students, pointing to the utilitarian nature of his theoretical framework, and consequent methodology.

REFERENCES

Aikenhead, G.S., (1996). Border crossing into the subculture of science. *Studies in Science Education, 27,* 1–52.

Aikenhead, G.S. (2001). Integrating western and aboriginal sciences: Cross-cultural science teaching. *Research in Science Education, 31,* 337–355.

Allchin, D. (1999). Values in science: An educational perspective. *Science and Education, 8*(1), 1–12.

Anderman, E.M., & Young, A.J. (1994). Motivation and strategy use in science: Individual differences and classroom effects. *Journal of Research in Science Teaching, 31*(8), 811–831.

Bartkus, K.R., & Stull, W.A. (2004). Research in cooperative education. In R.K. Coll & C. Eames, C. (Eds.), *International handbook for cooperative education: An international perspective of the theory, research and practice of work-integrated learning* (pp. 67–81). Boston: World Association for Cooperative Education.

Bennett, J., Rollnick, M., Green, G., & White, M. (2001). The development and use of an instrument to assess students' attitude to the study of chemistry. *International Journal of Science Education, 23*(8), 833–845.

Billingsley, B. (2000). Children's images of scientists: Stereotypes. In R.T. Cross, & Fensham P.J. (Eds.), *Science and the citizen* (pp. 79–83). Fitzroy, Australia: Arena.

Braunstein, L.A., & Loken, M.K. (2004). Benefits of cooperative education for employers. In R.K. Coll & C. Eames, C. (Eds.), *International handbook for cooperative education: An international perspective of the theory, research and practice of work-integrated learning* (pp. 237–246). Boston: World Association for Cooperative Education.

Butler, M.B. (1999). Factors associated with students' intentions to engage in science learning activities. *Journal of Research in Science Teaching, 36*(4), 455–473.

Chang, W., & Bell, B. (2002). Making content easier or adding more challenge in year one university physics? *Research in Science Education, 32,* 81–96.

Cobern, W.W. (1998). Science and a social constructivist view of the science education. In C.W. Cobern (Ed.), *Socio-cultural perspectives on science education: An international dialogue* (pp. 7–23). Dorechdt, The Netherlands: Kluwer.

Coll, R.K. (1996). The BSc(Technology): Responding to the challenges of the education marketplace. *Journal of Cooperative Education, 32*(1), 29–35.

Coll, R.K. (2004). Employers' views on the internationalization of cooperative education. *Journal of Cooperative Education, 38*(1), 35–44.

Coll, R.K. & Eames, C. (2004). Current issues in cooperative education. In R.K. Coll & C. Eames, C. (Eds.). (2004). *International handbook for cooperative education: An international perspective of the theory, research and practice of work-integrated learning* (pp. 271-282). Boston, MA: World Association for Cooperative Education.

Coll, R.K. & Eames C. (2008). Developing an understanding of higher education science and engineering learning communities. *Research in Science and Technological Education, 26*(3), 245–257.

Coll, R.K. & Taylor, N. (2004). Probing scientists' beliefs: How open-minded are modern scientists? *International Journal of Science Education, 26*(6), 757–778.

Coll, R.K., Lay, M., & Zegwaard, K. (2001). The influence of cooperative education on student self-efficacy towards practical science skills. *Journal of Cooperative Education, 36*(3), 58–72.

Coll, R.K., & Pinyonatthargarn, D., & Pramoolsook, I. (2003). The internationalization of cooperative education: A case study from Thailand. *Asia-Pacific Journal of Cooperative Education, 4*(2), 1–6.

Coll, R.K., Pinyonatthargarn, D., & Pramoolsook, I. (2004). Teaching technology and engineering in Thailand: Suranaree university of Technology as a model for cooperative education in Thailand. *Journal of Cooperative Education, 37*(2), 1–6.

Coll, R.K., Eames, C., Zegwaard, K., Hodges, D. (2002). How do we see ourselves: An Asia-Pacific regional perspective on cooperative education. In A. Zunaedi (Ed.), *Proceedings of the Fourth Asia-Pacific Conference on Cooperative Education* (pp. 1–5). Bandung: World Association for Cooperative Education.

Dalgety, J., & Coll, R.K. (2004). The influence of normative beliefs on students' enrolment choices. *Research in Science and Technological Education, 22*(1), 59–80.

Dalgety, J., & Coll, R.K. (2005). First-year tertiary chemistry students' attitude-towards-chemistry. *Canadian Journal of Science, Mathematics and Technology Education, 5*(1), 61–80.

Dawson, C., & O'Connor, P. (1991). Gender differences when choosing school subjects: Parental push and career pull. Some tentative hypotheses. *Research in Science Education, 21,* 55-64.

Dearing, R. (1997). Higher education in the learning society. *Report of the National Committee of Inquiry into Higher Education.* London: HMSO.

Dressler, S., & Keeling, A.E. (2004). Benefits of cooperative education for students. In R.K. Coll & C. Eames, C. (Eds.), *International handbook for cooperative education: An international perspective of the theory, research and practice of work-integrated learning* (pp. 217–236). Boston: World Association for Cooperative Education.

Duit, R. (Ed.). (2007). *Bibliography: Students' alternative frameworks and science education.* Kiel, Germany: University of Kiel.

Duit, R. (Ed.). (2009). *Bibliography: Students' alternative frameworks and science education.* Kiel, Germany: University of Kiel.

Eames, C. (2000). Learning in the workplace through cooperative education placements: Beginning a longitudinal qualitative study. *Journal of Cooperative Education, 35*(2–3), 76–83.

Eames, C. (2003). *Learning through cooperative education.* Unpublished PhD thesis, University of Waikato, Hamilton, New Zealand.

Eames, C., & Bell, B. (2005). Using sociocultural views of learning to investigate the enculturation of students into the scientific community through work placements. *Canadian Journal of Science, Mathematics and Technology Education, 5*(1), 153–169.

Eames, C., & Cates, C. (2004). Theories of learning in cooperative education. In R.K. Coll & C. Eames, C. (Eds.), *International handbook for cooperative education: An international perspective of the theory, research and practice of work-integrated learning* (pp. 37–47). Boston: World Association for Cooperative Education.

Eames, C., & Coll, R.K. (2006). Sociocultural views of learning: A useful way of looking at learning in cooperative education. *Journal of Cooperative Education & Internships, 40*(1), 1–13.

Fensham, P.J. (1980). Constraint and autonomy in Australian secondary science education. *Journal of Curriculum Studies, 12,* 186–206.

Fletcher, J. (1990). Self-esteem and cooperative education: A theoretical framework. *Journal of Cooperative Education, 26*(3), 41–55.

Fletcher, J. (1991). Filed experience and cooperative education: Similarities and differences. *Journal of Cooperative Education, 27*(2), 46–54.

Franks, P., & Blomqvist, O. (2004). The World Association for Cooperative Education: The global network that fosters work-integrated learning. In R.K. Coll & C. Eames, C. (Eds.), *International handbook for cooperative education: An international perspective of the theory, research and practice of work-integrated learning* (pp. 283–289). Boston: World Association for Cooperative Education.

Gardner, P.L. (1975). Attitudes to science: A review. *Studies in Science Education, 2,* 1–41.

Gibson, H.L., & Chase, C. (2002). Longitudinal impact of an inquiry-based science program on middle school students' attitudes toward science. *Science Education, 86,* 693–705.

Glasson, G.E., Mhango, N., Phiri, A., & Manier, M. (2010). Sustainability science education in Africa: Negotiating indigenous ways of living with nature in the third space. *International Journal of Science Education, 32*(1), 125–141.

Gogolin, L., & Swartz, F. (1992). A quantitative and qualitative inquiry into the attitude-toward-science of nonscience college students. *Journal of Research in Science Teaching, 29*(5), 487–504.

Groenewald, T. (2004). Towards a definition for cooperative education. In R.K. Coll & C. Eames, C. (Eds.), *International handbook for cooperative education: An international perspective of the theory, research and practice of work-integrated learning* (pp. 17–25). Boston, MA: World Association for Cooperative Education.

Good, R.G., Wandersee, J.H., & St Julien, J. (1993). Cautionary notes on the appeal of the new "ism" (constructivism) in science education. In K. Tobin (Ed.), *The practice of constructivism in science education* (pp. 71–87). Hillsdale, NJ: Lawrence Erlbaum.

Hanson, S.L. (1996). *Lost talent: Women in the sciences. Labor and Social Change Series.* Philadelphia: Temple University Press.

Jarvis, T., & Pell, T. (2001, March). *The effect of the challenger experience on children's attitudes to science.* Paper presented at the Annual Conference of the National Association for Research in Science Teaching, St Louis, MO.

Jarvis, T., & Pell, T, (2002a). Effect of the challenger experience on elementary children's attitudes to science. *Journal of Research in Science Teaching, 39*(10), 979–1000.

Jarvis, T., & Pell, T. (2002b, September). *Monitoring change in attitudes following a visit to an interactive simulation to space.* Paper presented at the annual conference of the British Educational Research Association, Exeter, UK.

Jones, M.G., Howe, A., & Rua, M.J. (2000). Gender differences in students' experiences, interests, and attitudes toward science and scientists. *Science Education, 84*(1), 18–192.

Jones, J., & Young, D. (1995). Perceptions of the relevance of mathematics and science: An Australian study. *Research in Science Education, 25*(1), 3–18.

Keane, M. (2008). Science education and worldview. *Cultural Studies of Science Education, 3*(3), 587-621.

Koballa,. T.R. Jr. (1990). Attitude. *Science Education, 74*(3), 369–381.

Krasilchik, M. (1990). The scientists: An experiment in science teaching. *International Journal of Science Education, 13*(3), 282–287.

Lave, J. (1991). Situated learning in communities of practice. In L. B. Resnick & J.M. Levine & S.D. Teasley (Eds.), *Shared cognition: Thinking as social practice, perspectives on socially shared cognition* (pp. 63–82). Washington, DC: American Psychological Association.

Lave, J. & Wenger, E. (1991). *Situated learning: Legitimate peripheral participation.* Cambridge, UK: Cambridge University Press.

Lederman, N.G. (1992). Students' and teachers' conceptions of the nature of science: A review of the research. *Journal of Research in Science Teaching, 29*(4), 331–359.

Lips, H.M. (1992). Gender- and science-related attitudes as predictors of college students' academic choices. *Journal of Vocational Behaviour, 40*(1), 62–81.

Lorsbach, A.W., & Jinks, J.L. (1999). Self-efficacy theory and learning environment research. *Learning Environment Research, 2,* 157-167.

Luzzo, D.A., Hasper, P., Albert, K.A., Bibby, M.A., & Martinelli, E.A. Jr. (1999). Effects of self-efficacy-enhancing interventions on the math/science self-efficacy and career interests, goals and actions of career undecided college students. *Journal of Counselling Psychology, 46*(2), 233–243.

McLaughlin, K. (1997). *Waterloo: The unconventional founding of an unconventional university.* Waterloo, ON: University of Waterloo.

Myers, R.E. III., & Fouts, J.T. (1992). A cluster analysis of high school science classroom environments and attitude toward science. *Journal of Research in Science Teaching, 29*(9), 929–937.

Nakhleh, M.B., Olles, J., & Malina, E. (2002). Learning chemistry in a laboratory environment. In J.K., Gilbert, O. De Jong, R. Justi, D.F. Treagust, & J.H. Van Driel (Eds.), *Chemical education: Towards research-based practice* (pp. 69–94). Dordrecht, The Netherlands: Kluwer.

Northeastern University. (2007). *Cooperative education.* Retrieved 05 June 2007, from http://www.northeastern.edu/coop/welcome2.html

Paku, L., Coll, R.K., & Zegwaard, K. (2003). Māori science and technology students' views of workplace support structures. In C. Gribble (Ed.), *Proceedings of the Seventh Annual New Zealand Conference on Cooperative Education* (pp. 24–34). Christchurch, New Zealand: NZACE.

Paku, L., Zegwaard, K., & Coll, R.K. (2003, August). *Enculturation of indigenous people into science and technology: An investigation from a sociocultural perspective.* Paper presented at the 13th World Conference on Cooperative Education, Rotterdam.

Perkins, D.N. (1997). Person-plus: A distributed view of thinking and learning. In G. Salomon (Ed.), *Distributed cognitions: Psychological and educational considerations* (pp. 88–110). Cambridge, UK: Cambridge University Press.

Pfundt, H., & Duit, R. (Eds.). (1994). *Bibliography: Students' alternative frameworks and science education.* Kiel, Germany: University of Kiel.

Pfundt, H., & Duit, R. (Eds.). (1996). *Bibliography: Students' alternative frameworks and science education*. Kiel, Germany: University of Kiel.

Pfundt, H., & Duit, R. (Eds.). (2000). *Bibliography: Students' alternative frameworks and science education*. Kiel, Germany: University of Kiel.

Piburn, M.D. (1993). If I were the teacher... Qualitative study of attitude toward science. *Science Education, 77*(4), 393–406.

Rennie, L.J., & Dunne, M. (1994). Gender, ethnicity, and students' perceptions about science and science-related careers in Fiji. *Science Education, 78*(3), 285–300.

Rogoff, B. (1995). Observing sociocultural activity on three planes: Participatory appropriation, guided participation and apprentice. In J.V. Wertsch, P. del Rio, & A. Alvarez (Eds.), *Sociocultural studies of mind* (pp. 139–164). Cambridge, UK: Cambridge University Press.

Ryder, J., Leach, J., & Driver, R. (1999). Undergraduate science students' images of science. *Journal of Research in Science Teaching, 36*(2), 201–219.

Ryder, K.G., & Wilson, J.W. (1987). *Cooperative education in a new era*. San Francisco: Jossey-Bass.

Schibeci, R.A. (1984). Attitudes to science: An update. *Studies in Science Education, 11*, 26–59.

Schibeci, R.A. (1986). Images of science and scientists and science education. *Science Education, 70*(2), 139-149.

Schibeci, R.A., & Riley, J.P. (1986). Influence of students' background and perceptions on science attitudes and achievement. *Journal of Research in Science Teaching, 23*(3), 177–187.

Smist, J.M., & Owen, S.V. (1994, April). *Explaining science self-efficacy*. Paper presented at the annual meeting of the American Educational Research Association. New Orleans, LA.

Sovilla, S.E., Varty, J.W. (2004). Cooperative education in the U.S., past and present: Some lessons learned. In R.K. Coll & C. Eames, C. (Eds.), *International handbook for cooperative education: An international perspective of the theory, research and practice of work-integrated learning* (pp. 3–16). Boston: World Association for Cooperative Education.

Speering, W., & Rennie, L. (1996). Students' perceptions about science: The impact of transition from primary to secondary school. *Research in Science Education, 26*(3), 283–298.

Talton, E.L., & Simpson, R.D. (1986). Relationships of attitudes toward self, family, and school with attitude-toward-science among adolescents. *Science Education, 70*(4), 365-374.

Talton, E.L., & Simpson, R.D. (1987). Relationships of attitude toward classroom environment with attitude toward and achievement in science among tenth grade biology students. *Journal of Research in Science Teaching, 24*(6), 507–525.

Thompson, J., & Soyibo, K. (2002). Effect of lecture, teacher demonstrations, discussion and practical work on 10th graders' attitudes to chemistry and understanding of electrolysis. *Research in Science and Technological Education, 20*(1), 25–35.

Tobin, K., & Tippins, D. (1993). Constructivism: A paradigm for the practice of science education. In K. Tobin (Ed.), *The practice of constructivism in science education* (pp. 3–21). Hillsdale, NJ: Lawrence Erlbaum.

Tymms, P. (1997). Science in primary schools: An investigation into differences in the attainment and attitudes of pupils across schools. *Research in Science and Technological Education, 15*(2), 149–159.

Vygotsky, L. (1978). *Mind in society: The development of higher psychological processes.* (M. Cole, V. John-Steiner, S. Scribner & E. Souberman, Trans.). Cambridge, MA: Harvard University Press.

Weisz, M., & Chapman, R. (2004). Benefits of cooperative education for educational institutions. In R.K. Coll & C. Eames, C. (Eds.), *International handbook for cooperative education: An international perspective of the theory, research and practice of work-integrated learning* (pp. 247–258). Boston: World Association for Cooperative Education.

Yang, E.M., Andre, T., & Whigham, M. (2001, March). *Self-Efficacy, liking, effort, expected grades, and perceived gender dominance for science and other school subjects among Czech and American students.* Paper presented at the annual meeting of the National Association for Research in Science Teaching. St Louis, MO.

PART II

SCIENCE ATTITUDE AND SOCIO-SCIENTIFIC ISSUES

CHAPTER 9

CULTURAL INFLUENCE ON ATTITUDES TOWARDS SCIENCE

Funda Ornek
University of Bahrain

ABSTRACT

Cultural influence plays an important role in the association of attitudes towards science, and this varies between countries because cultural context, including linguistic, social, political, economic, philosophical and religious aspects, determines and shapes attitudes towards science. Culture influence is the way we view the world. Learning, therefore, cannot be separated from its socio-cultural context (Vygotsky, 1987). Students carry their attitudes towards science into the classroom based upon their socio-cultural background influence. For example, Asian students choose science predominantly even though science-based careers are less economically profitable because of their families' important impact on their choice, and science-related choices carry a great deal of prestige within Asian cultures. Students within western cultures, on the other hand, are individualist and make attractive choices reflecting their personal enjoyment and skills to a greater degree. Moreover, in Turkish culture, science-related careers also have great prestige and have been perceived as very important for the country's development and therefore students' atti-

Attitude Research in Science Education, pages 241–261

tudes towards science are very positive even though their performance on the PISA is lower than average. As a result of different cultures holding different attitudes towards science and science-related careers, there is a great impact upon student uptake of science and science-related careers.

INTRODUCTION

Investigating students' attitudes towards choosing and studying science and science-related careers has been a fundamental focus in the science education community for more than 40 years since there is a significant decline in students' interest to pursue science-related careers. In addition, the numbers of students in middle and high schools turning away from science courses have increased (Atwater, Wiggins, & Gardner, 1995; Ayers & Price, 1985; Cannon & Simpson, 1985; Haladyna & Shaughnessy, 1982; Hill, Atwater, &Wiggins, 1995; Hofstein & Welch, 1984; Simpson & Oliver, 1990; Zacharia & Barton, 2004). This decrease will impinge on societies' economic and technological development because a knowledge-based economic future and advanced information society depends on an energized knowledge in science, math, and engineering. For this reason, contemporary science education emphasizes engaging students in science so that more positive attitudes towards science and science-related careers can be developed, because attitudes towards science may be considered to affect learning science (Lee & Erdogan, 2007). Accordingly, to increase enrollment in science courses, augment science achievement, and encourage young people, especially in primary, middle, and high school, to pursue science-related careers, more positive attitudes towards science can be an influencing aspect. For example, schools in most OECD (Organisation for Economic Co-operation and Development) countries do a reasonable job of transmitting science knowledge and skills, but they fail to engage students in science and science-related careers, which may be a hindrance for tomorrow's science-based societies (OECD, PISA, 2006). In other words, the decline in the number of science students has been raised as a vital concern with regard to each nation's economic future. This future can only be attained by increasing the uptake of science and science-related subjects. Otherwise, the negative attitudes towards science and science-related subjects pose a really serious threat to economic prosperity (De Boer, 2000).

How can more positive attitudes towards science and science-related careers be established? It is not that easy to answer this question because there are several external factors that can affect students' attitudes towards science and science-related careers in either a positive or negative way. On the other hand, a broad range of research has been conducted to focus on students' attitudes towards science and science-science related careers to understand the nature of the problem and to remediate the problem (Osborne, Simon, & Collins, 2003). These factors that influence students' attitudes towards science are mainly classroom (learning) environment, quality of teaching (teacher) in science courses, peer influence and parental influence on young students' science attitudes, the influence of gender on science attitudes, and the influence of culture and ethnicity on science attitudes (Collins, Michael, & Simon, 2006; Jinks & Morgan, 1999). This chapter seeks to emphasize the cultural influence on students' attitudes towards science and science-related careers.

CULTURAL INFLUENCE

Cultural influence has a significant effect on students' attitudes towards science and science-related careers such as medicine, forensic science, agriculture, and so forth among different countries. Cultural context has several components that determine and shape young people's attitudes towards science and even their achievements in science. These components consist of elements such as linguistics, social, political, economic, philosophical, and religious aspects. Evidently, cultural attributes are associated with organizational hierarchy, needs and beliefs of people, and even family influence. Furthermore, gender effect on science attitudes can be also considered a part of cultural attributes because socio-cultural influence plays an important role for males and females to establish either a positive or negative attitude towards science. For example, in Turkish culture, there is no difference between male and female uptake in any science careers or science-related professions. Thus, gender influence and cultural influence on students' attitudes towards science and science-related careers can be intertwined.

Culture persuades the way we view and experience the world. According to Vygotsky (1987), social context has a major influence on how we think and how we view the world. Culture constitutes the social context providing many of the views with which we see the world. Learning cannot be separated from its socio-cultural context. Therefore, socio-cultural background influences students' perception of science, achievement in science, and attitudes towards science and science-related careers or subjects. For example, Asian students choose science predominantly even though science-related careers are less economically profitable because of their families' important affect on their choice and science-related choices have a very great prestige within the Asian cultures. In addition, in Turkish culture, science-related careers also have great prestige and students' attitudes towards science is very positive even though students' performance on the PISA (*Programme for International Student Assessment*) is lower than average (OECD/PISA,

2006). Students within western cultures such as Europe and the U.S., on the other hand, are individualist and make attractive choices reflecting their personal enjoyment and skills to a greater degree. Moreover, in African cultures, the social-cultural background of students impacts negatively on students' attitudes towards science (Kesemang & Taiwo, 2002) because scientific knowledge is often seen as opposing traditional cultural beliefs such as taboos, omens, and witchcraft. Consequently, different cultures impact differently upon science and science-related selection.

In addition, socio-cultural factors have an impact on students' achievement in science as well as their attitudes towards science because they come to class with their naïve or intuitive views of science, not as blank boxes. Their background affects their learning of science and hence their attitudes towards science and science-related career choices.

Influence of Family on Attitudes towards Science

Families are the core of the culture. Families raise their children with their cultural essences so culture is embedded into children's lives. Therefore, family has a significant influence on students' choice whether these families are liberal families or authoritarian families. In both cases, families strongly influence students' attitudes towards science and science-related choices in either a positive or negative fashion. In many studies, it has been found that there are positive relationships between children's attitudes towards science and science-related choices and parents' attitudes towards science and science-related careers (Breakwell & Beardsell, 1992; Osborne et al., 2003; Talton & Simpson, 1985, 1986). For example, the majority of parents in Turkey have positive attitudes towards science and science-related choices and would support their children—whether they are male or female—if they chose, for example, to be a physics teacher as their profession (Erjem, 2000).

Parental influence can include a father's positive attitudes towards science, mother's support and encouragement of science, authoritarian or liberal family influence, or general family encouragement. As a part of the author's ongoing research to investigate science educators and math educators' views and observations regarding the influence of culture on students' attitudes towards science and science-related career choices, online interviews were conducted with several university professors in Science Education and several teachers (physics, science, mathematics) in Turkey and other countries to get their perspective with regard to Turkish culture's and other cultures' influence on students' science attitudes and their choice in science-related careers. The following quote from one professor in Amasya University explores the influence of parents on students' attitudes towards science:

I would like to think of a family as a culture because a child is immersed with the culture in the family—the way a child is raised, values, ethics, and religion. To say whether parents have positive or influence on their children's attitudes towards science, it is important that we need to look at the relationships between mother–child, father–child, mother–father and child. It is possible to have different kinds of family structures such as authoritarian, liberal, and careless families. In Turkish culture, father is mostly dominant in the family. Based on father's education level, children can have either positive or negative attitudes towards science and science-related careers. In our country, many studies showed that father's education level most influences 15-year-old students' attitudes towards science. (Professor in science education, Amasya University, Turkey, February 2010)

I think the culture is very important, and most of the influence comes from the home life. If you are growing up around parents who value science, and encourage it, then so will you. [That is] probably why private schools always have large science departments. So definitely the culture and home life, role models are massive factor. (Former science teacher in the UK, February 2010)

In addition, based on PISA 2006 results, in some European countries such as Ireland, France, Belgium, and Switzerland, students with a more advantaged socio-economic background were more likely to show a general interest in science. Another very significant feature of a student's background is if they have one parent in a science-related career. As a result, the parent's attitudes towards science and science-related careers are positively related to their children's attitudes toward science and science-related career choices (Kalender & Berberoglu, 2009).

Authoritarian families from certain cultures, coming mostly from the lower to moderate social status, have more positive attitudes towards science and science-related choices, whereas liberal families, who may have higher socio-economic status, have less positive attitudes towards science and science-related careers. Of course, this varies from culture to culture. For example, in western countries, children are raised more liberally regardless of whether they belong to lower socio-economic status or higher socio-economic status. Liberal families in general are in favor of individualistic decisions of their children, so they will not try to persuade their children's career choices. These parents support their children with regard to pursuit of careers that may be based on their enjoyment or abilities. Breakwell and Beardsell (1992) explored the finding that more liberal parents' children had less favorable attitudes towards science because of the parents' influence on their children. These parents support sex equality and encourage their children in political participation; in general, though, participation in political endeavors is more likely to lead to the social sciences. In addition, they pointed out one potential important reason why liberal families' children are more likely not to choose science or science-related choices—that liberalism may lead to children choosing more art-related careers. In these subjects, probably the effect of science in the environment is emphasized, and as a result, it may lead to students having negative attitudes towards science because science damages our environment—for example, nuclear power reactors. They do not see other ways that science can solve the problems related to the environment. Thus, although developed countries are at a very high level in terms of technology and science, their citizens are more likely to hold negative attitudes towards science. This leads to recruitment of professionals or students from abroad to work in the technology industry. For example, it is not surprising to see mostly Asian students (of course there are also students from other countries) in physics departments in U.S. universities, or Indian students in computer engineering and computer science departments. In other words, there have been a significant number of originally non American science students, scientists, and technologists participating in the development of science and technology of the U.S. (Haidar, 2002). Citizens may not feel or think that science is very important to the development of their country as they perceive their country is already at a very advanced technological level. On the other hand, there is a need to establish positive attitudes towards science and science-related careers and promote science nationwide, such as organizing Science, Math, and Technology festivals.

Among developing countries, Turkey is a country where science-related careers also have great prestige and have been perceived as very important for the country's development, and students' attitudes towards science are very positive even though students' performance on the PISA is lower than average, as shown in Figure 9.1. (In PISA 2006, more than 400,000 students from 57 countries took part. Science was the first focus, but it also included

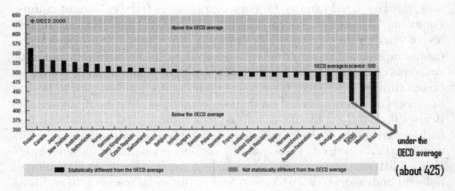

Figure 9.1 Performance (mean scores) on the science scale in PISA 2006 (OECD, PISA, 2006).

reading and mathematics. Data were on student, family, and institutional factors to explain difference in students' performance.)

As is seen in Figure 9.1, performance of Turkish students on the science scale in PISA (~425) is lower than OECD average in science (500). It is interesting to see a negative correlation between students' attitudes towards science and science-related careers and their performance on the science portion of PISA. From the point of view of a professor in science education in Turkey, he explains this amazing contrast as follows:

> In countries that have higher scores in PISA and TIMSS (Trends in International Mathematics and Science Study), GDP (gross domestic product) is high, technology is well-established, and there is no civil war. Therefore, in our country and in most countries in the world, based on the research results, it has been found that science and technology have an important place for students. In many countries, including Turkey, science and technology have been seen as an important factor for countries' development because a knowledge-based economic future and advanced information society depend on an energized knowledge of science, math and engineering;, science and technology can help us to find treatments for health problems such as cancer, AIDS, etc., and to find solutions for environmental problems such as air and noise pollution, nuclear waste, etc. Consequently, students are aware of and understanding of the importance of science, and students think advances in science and technology usually bring social benefits and important influences on everyone's life—whereas in developed countries, technology is developed and students have more likely less interest in science and technology. This shows that there is a satisfaction of technology. Along with this, developing and underdeveloped countries especially countries in Africa, people have incredible interest to technology and of course this is related to science. Therefore, in those countries people are more interested in science and technology. (Professor in science education, Amasya University, Turkey, February 2010)

As is explained by the professor, children's interest in science and technology is related to their understanding of the importance of science and technology. This awareness in general is given by parents. In addition, it is very important for families that their children study science and science-related careers in Turkey, because they think if their children succeed in science and math, that shows their children are very smart so that they are proud of their children. This factor pushes students to choose science or science-related careers, especially in authoritarian families. In society, science and science-related careers have prestige as well. The following quotes from a mathematics teacher and a professor in physics education in Turkey explore this:

> Parents always think that if their children succeed in Science, then they are certainly smart, successful, and outstanding because science especially phys-

ics is very difficult subject and it is not easy to be successful in science. (Math teacher in private teaching institute, Turkey, February, 2010)

In Turkey, science and science-related careers are popular especially medicine because it is easy to get a job after graduation and it also has important status in the society even though it takes a long time (at least 6 years if they do not continue to do TUS-Basic Medicine points in Examination of Specialization in Medicine) to become a doctor. Students should do well in science especially chemistry and biology, physics is of course required especially when students study on oncology or medical physics, but for basic knowledge of medicine the former ones are definitely required." (Professor in physics education, Turkey, February 2010)

In summary, the most significant impact upon students' attitudes towards science and science-related career choices correlates to that of the parents. This influence can be either positive or negative. For example, Asian students prefer to study science-related careers such as medicine-related studies (Modood, 1993; Osborne et al., 2003; Taylor, 1993) because Asian families greatly influence their children's career choices (Woodrow, 1996), much like Turkish families (Erdemir, 2004).

Influence of Religion on Attitudes Towards Science

Religion is a part of culture and has values, beliefs, and norms. It also has important influence on students' attitudes towards science and science-related career choices. These attitudes can be either positive or negative based on how much the society wants to merge religion with science. In addition, it depends on the interpretation of the holy books in different religions such Qur'an in Islam or Bible in Christianity. So, people's worldviews vary with their religious background. The author will consider two religions to explain the major influences that religion may have on people's ways of thinking about science and science-related career choices. Of course, there are other religions that may impact on people's ways of thinking about science and science-related career professions. It is unfortunate that science and religion are perceived as two different entities (Guessoum, 2009) and there has always been a conflict between the two, whereas science and religion could work together in harmony. For example, Islam and science had worked in harmony until Ottoman Empire was destroyed (early 1900s). Hüseyin Hilmi Işık (2001), a scientist (Pharmacist and Chemical Engineer who synthesized and determined a formula for the ester "phenylcyannitromethan-methyl") was also an Islamic scholar, stated "... *Islam was misrepresented as hostile towards knowledge, science, as hostile towards knowledge, science and bravery, while, in fact, it is the protector of such things, encouraging every kind of progress and improvement*" (p. 12). Because of misrepresentation of Islam, as stated by Işık (2001), countries with Muslim populations lost some of their understanding contemporary knowledge in science, which is one of the most important requirements of 21st century. To develop and maintain a literate citizenry, science is a fundamental requirement. (Tobin, 2010). Within the Muslim world, it has proved difficult to sustain the accomplishments in science, math, and engineering and even in military defense, as there was been a prevalence to compartmentalize, and hence separate science and religion. This has caused people from these countries to fall behind in the knowledge of science, leading to a major obstacle in the development of said countries. A secondary global issue, as a result of this, is the negative image portrayed of the religion, as Islam is often labeled as the culprit (Işık, 2001). Thus, students may have negative attitudes towards science and science-related careers choices such as genetics, medicine, and physics. He stated: "... *we have to abolish this black curtain and get rid of the oriental religion...*" (p. 13). Then, positive attitudes towards science and science-related career choices can be granted. Even in history, the most famous scientists such as Newton, Kepler, and Leibniz were very religious and followed their religions tightly (Işık, 2001).

For example, Isaac Newton (1642–1727), who was physicist, mathematician, astronomer, philosopher, and theologian, was one of the most prominent people in the history. Besides his scientific fame, Newton, it was also noteworthy that he studied the Holy Bible. Newton's best-known discoveries were the laws of motion and universal gravitation. Newton's laws of motion are composed of three laws: 1) Newton's firs law of motion states: *"an object moves in a straight line and at constant speed except to the extent that it interacts with other objects."* (Chabay & Sherwood, 2002, p. 19). It is also known as the law of inertia. 2) Newton's second law of motion states that

$$\frac{d\vec{p}}{dt} = \vec{F}_{net} = \frac{d(m\vec{v})}{dt} = m\vec{a}.$$

3) Newton's third law of motion states that the force that object A exerts on object B is equal and opposite to the force that object B exerts on object A. It is also called "reciprocity." Newton's law of universal gravitation states: *"the force F_g acts along a line connecting the two objects, and its magnitude is proportional to the mass m_1 of the first object and to the mass m_2 of the second object, and inversely proportional to the square of the distance r between two objects."*(see Figure 9.2) (Chabay & Sherwood, 2002, p. 38).

After his discoveries, he said; *"Gravity explains the motions of the planet, but it cannot explain who set the planets in motion. God governs all things and knows all that is or can be done."* (Wikipedia). As is explained, science and religion were exactly parallel with each other in those days. Science and religion as parallel lines should embrace each other.

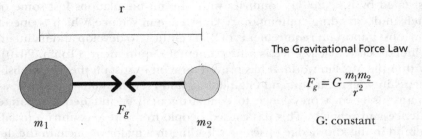

Figure 9.2 The gravitational force.

In Islam (mostly Arab culture is considered even though it does not relate to only Arab culture), modern science is not welcomed very well even in the Western world. Christianity has not embraced some of modern science, such as cloning especially human cloning (Haidar, 2002) and theory of evolution. In Christianity, especially the Catholic Church apposes all form of cloning because the concept of cloning is not consistent with the Holy Bible's view of human life. It was stated: "*Human cloning is intrinsically illicit. . . . From the ethical point of view, so-called therapeutic cloning is even more serious. To create embryos with the intention of destroying them, even with the intention of helping the sick, is completely incompatible with human dignity.*" (Pullella & Lewis, 2008). On the other hand, prohibition of cloning will limit scientific progress. According to scientists, to ban cloning for especially medical research may limit scientific progress. Moreover, researcher and scientists believe; "*stem cells taken through cloning are useful for treating several diseases, such as diabetes and Alzheimer's, because the resulting tissue could be genetically matched to a patient's body, avoiding immunological rejection.*" (Brainard, 2003). Therefore, this banning due to the influence of religion will impact people's attitudes towards science and science-related careers.

Here is one example from the author's research about teacher candidates' views of Nature of Science conducted with Arab science teacher candidates while they were doing their PGDE-Post Diploma Graduate Education. The results were presented in the CRPP conference-Centre for research in Pedagogy and Practice in Singapore in 2009. The following quotes demonstrate that Science teacher candidates' attitudes towards science especially genetics are not positive because they think it is against the teaching of Islam.

"I agree that science is infused with social and cultural values because science is related to cultural, and social and philosophical assumption, we are Muslim and the knowledge required in Islam but we must take in mind the legitimacy term for example genetic science is important in science but we cannot use all application like cloning in human or animals because it is interference of God's creation."(Science teacher candidate 1, Fall 2008)

"Science should be universal but some time it affected by the culture for example non-Islamic world can use science to make cloning but it's not acceptable with Islamic world." (Science teacher candidate 2, Fall 2008)

On the contrary; Islam is a religion which encourages all aspects of knowledge, all aspects of science and all aspects of experimentation (Işık, 2001). Therefore, one would expect Muslims to be in favor of science and encourage their fellows to carry out experiments in science. Işık (2001) affirmed; *"Islam is a dynamic religion commanding us to study science, experiment and to do positive work."* In many places of the Islamic holy book, Qur'an, human beings are commanded to see and observe nature, which includes all creatures, living and lifeless beings. Here is an example from the days of the Prophet (571–632) which illustrates how important it is to do experimentation in Islam.

"One day his As-hâb-ikirâm 'alaihim-ur-ridwân' (friends of the Prophet) asked our Prophet, Some of us who have been to Yemen saw that they budded the date trees in a different way and got better dates. Shall we bud our trees in Medina as our fathers have been doing or as we have seen them do in Yemen, thus getting better and more plentiful dates?" Rasûlullah could have answered them, "Wait a bit! When Hadrat Jabrâil (Gabriel) comes, I will ask him and tell you what I learn," or "I must think for a while; when Allahu ta'âlâ lets my heart know the truth, I will tell you." He didn't. Instead, he said, "Try it! Bud some of the trees with your father's method, and others with the method you saw being used in Yemen! Then always use the method which gives better dates!" In other words, he commanded us to experiment and to rely on experimentation, which is the basis of science. He could have learned it from the angel or no doubt, it might have materialized in his blessed heart. But he pointed out that all over the world Muslims who will exist until the end of the world should rely on experimentation and science."(Işık, 2001, p36).

Because of misunderstanding and the misrepresentation of some concepts such as cloning and evolution in science, some families in Arab culture have negative attitudes towards science. As a result of their influence, their children also foster these negative attitudes towards science. When, in fact, cloning is not understood well enough to make informed comments on it, as is seen, in above excerpts from some Arab science teacher candidates. The findings are sometimes conveyed in a way that people get confused and refuse the new invention because of conflict with their religious values that they believe in. As can be recalled, there was a very intriguing headline regarding cloning saying that people created a sheep and they will create a human in media and newspapers. Whereas cloning is not creating a new thing, it is to produce of copies of DNA fragments (molecular cloning), cells (cell cloning), or organisms as explained in Wikipedia encyclopedia.

Prof. Ramazan Ayvalli in Faculty of Theology (Educated in Egypt and Turkey) explained cloning from a religious point of view. Fertilizing an egg with sperm and providing suitable environment for this fertilized egg is not the creation of something from nothing. It is like taking a picture of a person or an object or making a copy of something. Human being can make changes to already existing things. Therefore, it is not possible to create a human or even a fly. Human exists with their soul. The body of a human is like a carpenter's tools. Even if one's all organs are replaced with different people's organs, the person's mind, thinking, intelligence, and knowledge are not going to be changed. In other words, carpenter's old tools are replaced with new tools. As changing tools, knowledge and talent of carpenter do not change. For example, if a blind person has eye transplantation, then s/he may see. Even if one's blood, brain, heart are changed, this change does affect their thinking. Healthy organ functions well and human means soul. Theoretically, humans can be copied, but souls cannot be copied. Soul is not related to genetic makeup. Each living being has a different soul. Moreover, souls of identical twins are different.

Even if a famous person is copied; the baby from cloning will have different knowledge, thinking, intelligence, and abilities from that famous person. Let's say you want to have many scientists like Einstein to have his brain to solve many problems regarding physics, and you have him genetically copied. The babies will have same the appearance as Einstein but not his thinking, his intelligence, knowledge, and brain functions because the new babies will have their own psychological conditions and they will have different life experiences to him.

As scientists explained, either mother or father is not a part of cloning. If genetic character comes from the mother, then the mother's chromosomes are mixture of her own parents' chromosomes. Therefore, foetus carries her/his grandmother's and grandfather's chromosomes and not only his/her own mother. A problem with regards, this causes disposition for a race. For this reason, human cloning can be harmful.

Prof Ayvalli stated that, in some newspapers and media, there are headlines such as 'human created a sheep' or 'human is creating a human'. Cloning is not equivalent to creating as creating is making something from nothing. In cloning, genetic substances inside a cell which is created by Allah is used. These substances are injected into the mother's egg. The soul is given by Allah. This cannot be called creating. It is like making a sweet by mixing flour, sugar, and oil. Flour, sugar, and oil cannot be created from nothing. However, a new product can be produced by using available substances. In cloning, only physical characteristics are copied. For example, identical twins have similar DNA characteristics, that's; their physical characteristics are similar, but the souls are different.

It is beneficial to rear animals to produce better quality meat, milk, eggs and is not against Islam. Moreover, married couples who cannot conceive babies in naturally, genetic material from the father would be injected into the mother's egg which would then be implanted into mother's womb to grow. If it is not harmful for human health to eat cloned animals such as sheep, it is not right to be against animal cloning.

Italian Prof. Dr. Severino Antinori who works in cloning stated that cloning is not like photocopying. We do not produce the same people. Copying of a body can be done, but the person's psychological conditions cannot be copied. Therefore, a person with the same soul cannot be copied.

As is explained by Prof. Severino Antibori and Prof. Ramazan Ayvali, there is no conflict between religion and science because cloning is not creating something that does not already exist. So understanding that science and Islam can work in harmony should have a positive influence on both families' and children's attitudes towards science, as they will see that Islam encourages human beings to study science, do experimentation for the purpose of human well-being. The religion emphasizes that people should trust in science and experimentation to obtain the best results for society's prosperity.

One of the reasons that Turkish students choose science or science-related careers to study is that their religion, Islam, influences their choices. They follow what Islam says about the science and the importance of science in our life. Without science, the country cannot be developed and provide the necessary social benefits to its citizens. They believe that they lost knowledge of science and need to seek knowledge of science and bring it to their countries. In fact, students follow these two hadithes of the Prophet, which state: "*Hikmat (that is science and art) is the lost property of the Muslims. Let him take it wherever he finds it!*" and "*Seek knowledge of science even in China!*" (Işık, 2001). (An hadith is the words and deeds of the Islamic prophet). The following quotes from a physics teacher and a science teacher in Turkey reinforce the above statement more.

> "Even though the influence of culture on students' attitudes towards science has been decreasing in globalization world and we are establishing same thinking with the western world, our religion is still taking a part of students' life as Turkish culture. Therefore, students still follow the hadith of our prophet saying "Seek knowledge of science even in China!"and they chose to study science. Moreover, they leave their countries to study broad and return to their country to serve to the country to develop our country. Their goal is to produce resources for Turkey. These students are mostly from a low socio-economic level and very determined. They are mostly a grandchild of martyr in the war and they are aware that their country cannot be protected by weapon instead,

knowledge-based economic development. S/he knows that knowledge-based economic future and advanced information society depends on an energized knowledge in science, math, technology, and engineering." (Science teacher, Turkey, February 2010)

"Religion of Islam is the major religion in Turkey and science in Islam is very important because Islam is against to not being educated. As known, religion affects culture of a country. Religion and culture are very much related, so I can say that people have positive attitudes towards science through an effect of religion. Families always want to raise children in a way to serve to their country to develop the country. In addition, families want their children to provide knowledge-based economic future and advanced information society for the country. Therefore, Turkish culture has positive influence on students' and families' attitudes towards science."(Physics teacher, Turkey, February, 2010)

In summary, religion can have a potential positive or negative effect on people's attitudes towards science and science-related careers. As most researchers point out, students, who consider science and religion in conflict with each other in terms of seeking knowledge, do not have any improvement on the nature of science views (Mugaloglu & Bayram, 2009; Abd-el-Khalick & Akerson, 2004; Roth & Alexander, 1997). Religious values of students or even professionals is in fact a significant predictor towards students' attitudes with regards science and they can establish negative attitudes towards science because they think some topics or concepts are not appropriate for religious values. So there is always a potential interference between students' religious values and their attitudes towards science even though their understanding of these controversial concepts has flaws as explained above regarding cloning.

INFLUENCE OF SUPERSTITIONS ON ATTITUDES TOWARDS SCIENCE

Superstitions are a part of culture and have a significant effect on people's lives including personal and professional lives. Superstition, as is defined in Wikipedia Encyclopedia, is a credulous belief and is not based on reason, knowledge, experience, or observation. It is mostly known as 'folk beliefs' or 'old wives tales'. It is in general related to beliefs and practices such as luck (bad or good), prophecy, and spiritual beings. There is no rational behind this type of beliefs that can affect future events as the superstation would have and believe (Foster & Kokko, 2009). For example, here are two common examples regarding superstitions: If a black cat crosses a parson's pathway, this is bad luck or it is good luck when a ladybug lands on a person.

In addition, some superstitions come from religious practices and people continue to observe these superstitions even though they do not adhere to any religion. In general, these practices lost their original meanings in this transition. In some cases, the practices are adapted to the current religion the people practice. For instance, pagan symbols to protect against evil were replaced with the Christian cross during the Christianizing of Europe.

Superstitions are socio-cultural factors that play a significant role in science learning of children and their attitudes towards science in mostly non-western cultures such as African culture. Of course, many superstitions take place in other cultures but not to the degree of impact which superstitions have as in African culture. For example, there is cultural conflict between the scientific culture and African, African-American, Native American cultures (Lemke, 2001; Allen & Crawley, 1998; Aikenhead, 1996; Cobern, 1996; Costa, 1995; Barba, 1993). Moreover superstitions took place especially in rural parts of Turkey before 1990s in Turkish culture. It was unfortunate that these beliefs were embedded in the religion so these beliefs affected students' science learning, students' attitudes towards science, and their science-related career choices. A physics teacher in Turkey says,

> "because unfortunately, students who had low socio-economic background in rural areas had nonscientific beliefs as a part of the religion, their attitudes towards science and science learning were negative so that they did not choose to study science for their further endeavors."

In African culture, superstitions impact negatively on students' learning of science (Kesemang & Taiwo, 2002). Several African educators found out that the impact of socio-cultural factors impact African children's attitudes towards science and even their achievement in science (Kesemang & Taiwo, 2002; Ogunniyi, 1988; Jegede & Okebukola, 1991). Researchers and educators point out that children's socio-cultural background may be a barrier to their productive learning of science. It will definitely affect their thinking skills and process. For example, as is shown in Kesemang & Taiwo's study regarding Botswana culture, in most African culture, children do not believe that people die of natural causes such as cancer, or AIDS. They believe that a person dies because of witchcraft. Children believe that the death of person is due to bewitching by a relation, an enemy of the person's family, a witch doctor, or a traditional medicine person. As an example from Nigerian culture, people consider the chameleon to be evil. Another very interesting example relates to lightning; when a person is struck by lightning, it is believed that that a witch doctor or an evil medicine man or woman is responsible for that person's death. Therefore, the child has been raised with the influence of non-rationale beliefs surrounding them including beliefs about natural phenomena such as lightning and thunder. Children's thinking is shaped and dominated by these beliefs.

Moreover, children in African culture believe that the products of science and technology such as aeroplanes, electricity, and so forth are the white man's magic. Children raised in African culture are affected by these beliefs so that their attitudes towards science and their achievement in science learning are negatively influenced (Kesemang & Taiwo, 2002). This does not mean that all people in African culture have only negative attitudes and their achievement in science is low. There are many African people who are majors in science or scientists, but they may retain their cultural knowledge in synchronization with their scientific knowledge. These people may be able to study modern ecology view the chameleon is evil (Cobern, 1996). However, these people often hide their beliefs in cultural knowledge when they are with other people who are not raised in the African culture. Here is a quotation from an African professor in educational psychology who was educated in both Africa and Canada and taught in Africa and Bahrain, he explores the influence of superstitious beliefs including religion on students' attitudes towards science.

> "...Causation system among modern African students influenced by both the Western scientific paradigm and African indigenous world views and in addition influenced by Christianization and Islamization. Causation among students in African is multi-dimensional and it would be difficult for me to point at any single factor without research to see which of these variables factors influence African students' attitudes towards learning science. From ordinary observation need also to consider the level of acculturation, education and religious influence in particular African communities. When you unpack all these causation factors, you might find for examples that typically rural students (with strong rural background so called SRB) love to learn science but tend to be influenced by superstitious beliefs that conflict with scientific reasoning- Then there certain religious sects with strong beliefs conflicting with scientific knowledge. (Professor in Educational Psycology, Bahrain, February 2010).

As a result, cultural beliefs-taboos, omens, or witchcrafts held by students affect negatively on their science attitudes and science achievement. Here are some examples for taboos, omens, and witchcrafts. 'Water or fire should not be brought into (one's) compound at night' is an example of taboos. 'When cocks fight, it is a sign that there will be a visitor to one's compound later on that day' is an example of omens. Finally, 'When a person dies of AIDS-related ailments, s/he is generally said to have been 'bewitched' during burial rites' is an example of witchcraft. For example, in Jegede & Okebukola (1991) study, they found that traditional African beliefs-taboos, omens influence students' science learning. The conflict between traditional culture and scientific culture causes that students have unfavorable attitudes towards science (Jegede & Fraser, 1989). However, it is important to remember that African students can hold a traditional African view as well as a scientific view and they can have positive attitudes towards science and science learning.

SUMMARY

This chapter has sought to provide a review of cultural influence on students' attitudes towards science and science-related career choices including some interviews with experts in science and mathematics education of ongoing research of the author. As is elucidated in this chapter, cultural influence on students' attitudes towards science and science-related career choices plays a significant role and affect their future choices. The influence of culture is explained in three sub-categories which are the influence of families, the influence of religion, and the influence of superstitions on students' attitudes towards science and science-related career choices. Their influences can be either positive or negative based on the culture of societies. As the author pointed out, all research done so far has shown that different societies hold different perspectives on science and the value of science-related careers such as genetic engineering, medicine, and agriculture. For example, Asian students prefer to study for degrees in science related careers such as medicine-related studies, engineering, or even mathematics probably because of parental involvement. On the contrary, Afro-Caribbean students do not choose to study science or science-related studies; instead they prefer to pursue degrees in the social sciences (Osborne et al., 2003) such as history, economics. Many students' attitudes towards science are influenced by their parents because they want to please their parents even though students may not be fond of science. Therefore, parental involvement has a major role in influencing their children's attitudes towards science, advising, and guiding their children's career choices (Collins, Michael, & Simon, 2006).

Religion also has a big influence on students' attitudes towards science and science-related choices because mostly the message of the religion is conveyed to society in an obstructive way. This is especially happening in Islam because discussions of science and Islam continue and this prevents both students and families from having positive attitudes towards science and studying science or science-related career degrees as explained by the author. For example, it is very obvious that in today's world, the majority Muslim countries' contributions to science are negligible (Edis, 2009). On the other hand, Muslim scientists in non-Muslim countries have significant contributions to science. Of course, Islam's influence on students' attitudes towards science is positive in some other countries such as Turkey.

Cultural beliefs including superstitions-taboos, omens, and witchcrafts impact negatively upon students' science learning, students' science attitudes, and their achievement in science because students come to the science class with cultural beliefs pre-occupied in their minds which can be opposite to what is known to be scientifically correct knowledge. Since their cultural beliefs are opposite to scientifically correct knowledge, they may not choose to study science-related careers. Therefore, there is an essential need to replace students' unscientific knowledge that they hold as a result of their cultural beliefs such as taboos, omens, and witchcrafts with scientifically correct concepts. For example, the unscientific concept that they have of 'lightning can be sent by a witch doctor to strike other people' should be replaced by the scientific explanation which is that it is briefly enormous electrical discharge caused because positive and negative charges are imbalanced. To attract students to science and science-related careers, traditional beliefs culture and scientific culture should be linked in a way that students learn science and their science attitudes are influenced positively.

In conclusion, cultural difference including family background, religion, and superstitions-taboos, omens, and witchcrafts is a significant aspect of many students' attitudes towards science and science-related career choices. In order to improve science education, attract, and retain more students to science and science-related subjects, the science educators, religious scholars, families, and policy makers need to identify successful ways that allow students to practice science and allow cultural border-crossing for students.

ACKNOWLEDGEMENT

The author would like to thank Samina Shujaat (former science teacher) for her thoughtful comments on the draft version of this chapter.

REFERENCES

Abd-El-Khalick, F., & Akerson, V. (2004). Learning as conceptual change: Factors mediating the development of preservice elementary teachers' views of nature of science [Electronic version]. *Science Education, 88(5),* 785–810.

Aikenhead, G.S. (1996). Science education: Border crossing into the subculture of science. *Studies in Science Education, 27,* 1–52.

Allen, N.J., & Crawley, F.E. (1998). Voices from the bridge: Worldview conflicts of Kickapoo students of science. *Journal of Research in Science Teaching, 35,* 111–132.

Atwater, M., Wiggins, J., & Gardner, C. (1995). A study of urban middle-school students with high and low attitudes toward science. *Journal of Research in Science Teaching, 32,* 665–677.

Ayers, B., & Price, O. (1985). Children's attitudes toward science. *School Science and Mathematics, 75,* 457–460.

Barba, R.H. (1993). A study of culturally syntonic variables in the bilingual/bicultural science classroom. *Journal of Research in Science Teaching, 30,* 1053–1071.

Brainard, J. (2003). Cloning debate moves to the states. *The Chronicle of Higher Education, 49*(29), A22–A23.

Breakwell, G. M. & Beardsell, S. (1992). Gender, parental and peer influences upon science attitudes and activities [Electronic version]. *Public Understanding of Science, 1,* 183–197.

Cannon, R., & Simpson, R. (1985). Relationships among attitude, motivation, and achievement of ability grouped, seventh grade, life science students. *Science Education, 69,* 121–138.

Chabay, R. W. & Sherwood, B. A. (2002). *Vol I: Matter and interactions: Modern mechanics.* New York: John Wiley & Sons, Inc.

Cobern, W.W. (1996). Worldview theory and conceptual change in science education. *Science Education, 80,* 579–610.

Cobern, W. W. (1996). *Traditional culture and science education in Africa: Merely language games?* Paper presented at the meeting for Traditional Culture, Science and Technology, and Development: Toward a New Literacy for Science and Technology Tokyo Institute of Technology, Meguro-ku, Tokyo, Japan.

Collins, S., Michael, R., & Simon, S. (2006). *A literature review of research conducted on young people's attitudes to science education and biomedical science* (The Wellcome Trust). London, UK: Institute of Education, University of London.

Costa, V.B. (1995). When science is another world: Relationships between worlds of family, friends, school, and science. *Science Education, 79,* 313–333.

De Boer, G. (2000). Scientific literacy: Another look at its historical and contemporary meanings and its relationship to science education reform [Electronic version]. *Journal of Research in Science Teaching, 37,* 582–601.

Edis, T. (2009). Modern science and conservative Islam: An uneasy relationship [Electronic version]. *Science and Education, 18,* 885–903.

Erdemir, N. (in press). Fizik öğretmeni adaylarının bölümü tercih nedenleri ve mekanik başari düzeylerine etkisi. *Erzincan Egitim Fakultesi Dergisi.*

Erdemir, N. (2004). *An identification of physics student teachers' changing of successes and attitudes in their education processes.* Unpublished Doctoral Dissertation, Science Institute, Black Sea Technical University, Turkey.

Erjem, Y. (2000). Öğretmenlik mesleğine yönelmede ailenin İşlevi, öğretmenlik meslek bilgisi Programına katılan Öğrenciler Üzerine bir araştırma, Çukurova Üniversitesi eğitim Fakültesi Dergisi, Cild: 2 Sayı 19.

Foster, K. R. & Kokko, H. (2009). The evolution of superstitious and superstition-like behavior [Electronic version]. *Proceedings of the Royal Society B: Biological Sciences, 276,* 31–37.

Guessoum, N. (2009). Science, religion, and the quest for knowledge and truth: An Islamic perspective [Electronic version]. *Cultural Studies of Science Education, 5,* 55–69.

Haidar, A. H. (2002). Emirates secondary school science teachers' perspectives on the nexus between modern science and Arab culture [Electronic version]. *International Journal of Science Education,* 24(6), 611–626.

Haladyna, T., & Shaughnessy, J. (1982). Attitudes toward science: A quantitative synthesis. *Science Education, 66,* 547–563.

Hill, G., Atwater, M., & Wiggins, J. (1995). Attitudes toward science of urban seventh-grade life science students over time, and the relationship to future plans, family, teacher, curriculum, and school. *Urban Education, 30*(1), 71–92.

Hofstein, A., & Welch, W. W. (1984). The stability of attitudes towards science between junior and senior high school. *Research in Science and Technological Education, 2,* 131–138.

Hofstein, A., Maoz, S., & Rishpon, M. (1990). Attitudes toward school science: A comparison of participants and nonparticipants in extracurricular science activities. *School, Science and Mathematics, 90*(1), 13–22.

Işık, H. H. (2001). *Endless bliss. First fascicle* (8th, 10th, and 12th ed.). Istanbul: Hakikatkitabevi: Waqf Ikhlas Publications.

Jegede, O. J. & Okebukola, P. A. O. (1991). The relationship between African traditional cosmology and students' acquation of science process skills. *International Journal of Science Education, 13,* 37–47.

Jegede, O. J. & Fraser, B. (1989). Influence of socio-cultural factors on secondary school students' attitude towards science [Electronic version]. *Research in Science Education, 19,* 155–164.

Jinks, J. L. & Morgan, V. (1999). Children's perceived academic self-efficacy: An inventory scale. *Clearing House, 72,* 224–230

Kalender, I. & Berberoglu, G. (2009). An assessment of factors related to science achievement of Turkish students [Electronic version]. *International Journal of Science Education, 31*(10), 1379–1394.

Kesemang, M. E. E. & Taiwo, A. A. (2002). The correlates of the socio-cultural background of Botswana junior secondary school students with their attitudes towards and achievements in science [Electronic version]. *International Journal of Science Education, 24*(9), 919–940.

Lee, M. K., & Erdogan, I. (2007). The effect of science-technology-society teaching on students' attitudes towards science and certain aspects of creativity [Electronic version]. *International Journal of Science Education,* 29(11), 1315–1327.

Lemke, J. L. (2001). Articulating communities: Sociocultural perspectives on science education. *Journal of Research in Science Teaching, 38*(3), 296–316.

Modood, T. (1993). The number of ethnic minority students in higher education. Some ground for optimism. *Oxford Review of Education, 19,* 167–182.

Mugaloglu, E.Z. & Bayram, H. (2009). Do religious values of prospective teachers affect their attitudes toward science teaching? [Electronic version]. *Turkish Journal of Science Education, 6*(3), 91–98.

OECD, PISA. (2006). Retrieved March 28, 2009 from http://www.oecd.org/docum ent/40/0,3343,en_2649_34319_41689640_1_1_1_37417,00.html

Ogunniyi, M. B. (1988). Adapting western science to traditional African culture. *International Journal of Science Education, 10,* 1–9.

Ornek, F. (2009). *Bahrain teachers candidates' views of nature of science: A phenomenographic study.* Paper presented at the 3rd Redesigning Pedagogy International Conference (CRPP), Singapore.

Osborne, J., Simon, S., & Collins, S. (2003). Attitudes towards science: A review of the literature and its applications [Electronic version]. *International Journal of Science Education, 25*(9), 1049–1079.

Pullella, P. & Lewis, C. (2008). Vatican document denounces in vitro; Bioethics; Embryonic stem-cell research seen as immoral. *World,* A19

Roth, W., & Alexander, T. (1997). The interaction of students' scientific and religious discourses: Two case studies [Electronic version]. *International Journal of Science Education, 19*(2), 125–146.

Simpson, D., & Oliver, S. (1990). A summary of major influences on attitude toward and achievement in science among adolescent students. *Science Education, 74,* 1–18.

Talton, E. L. & Simpson, R. D. (1985). Relationships between peer and individual attitudes toward science among adolescent students. *Science Education, 69,* 19–24.

Talton, E. L. & Simpson, R. D. (1986). Relationship of attitudes toward self, family, and school with attitude toward science among adolescents. *Science Education, 70,* 365–374.

Taylor, P. (1993). Minority ethnic groups and gender access in higher education. *New Community, 19,* 425–440.

Tobin, K. (2010). Issues of our time: Science, religion, and literacy [Electronic version]. *Cultural Studies of Science Education, 5,* 1–4.

Vygotsky, S. L. (1987). Thinking and speech. In R.W. Riber & A.S. Carton (Eds.), *The collected works of S.L. Vygotsky, Volume 1: Problems of general psychology.* New York: Plenum.

Woodrow, D. (1996). Cultural inclinations towards studying science and mathematics. *New Community, 22,* 23–8.

Zacharia, Z. & Barton, A. C. (2004). Urban middle-school students' attitudes toward a defined science [Electronic version]. *Science Education, 88*(2), 197–222.

Wikipedia, the free encyclopedia, retrieved February 20, 2010 from http://en.wikipedia.org/wiki/Cloning

Wikipedia, the free encyclopedia, retrieved February 20, 2010 from http://en.wikipedia.org/wiki/Superstition

Wikipedia, the free encyclopedia, retrieved February 20, 2010 from http://en.wikipedia.org/wiki/Isaac_Newton

CHAPTER 10

STUDENT ATTITUDES TOWARD SCIENTISTS

Anita Welch
North Dakota State University

Douglas Huffman
University of Kansas

ABSTRACT

Students tend to have stereotypical views of scientists and tend to see science as a unique activity performed by special people. This stereotypical view of scientists is associated with a negative attitude toward learning science. In this chapter, we examine students' attitudes toward scientists from both a theoretical and practical perspective. Students' attitudes towards scientists are explored through a study on the impact of an after-school robotic competition. Secondary students participated in a robotics competition where they worked with scientists to design and build a robot to perform a specific task. Students' attitudes toward the scientific process and attitudes toward scientists were measured using the Test of Science Related Attitudes (TOSRA). Students who participated in the robotics competition were compared to students in the same schools who did not participate in the robotics competition. Results indicated that students who participated in the robotic competition had a more positive attitude toward scientists and science. The robotics competition ap-

Attitude Research in Science Education, pages 263–279

pears to have helped students see science as a normal activity, and it appears to have helped students see scientists as normal people, although students still retained the view of scientists as stereotypical smart people referred to as "nerds." The theoretical and practical implications of results for teaching science are discussed.

INTRODUCTION

One of the key challenges facing the field of science education is recruiting, educating, and retaining students in the field of the sciences, technology, engineering, and mathematics. Within the next decade, "the number of individuals with science and engineering degrees reaching typical retirement age is expected to triple" (National Science Foundation, 2002, p. 31). In 1999, among 3,540,800 persons employed in science and engineering occupations, only 1,032,100 had Master's degrees and 484,100 had earned Doctorate degrees (Wilkinson, 2002). In a report from the Merrill Advanced Studies Center, Ortega states, "The fundamental problem is the declining percentage of students in science, technology, engineering, and mathematics (STEM) graduate programs, especially at the doctoral level" (2003, p. 12). In addition, student interest and attitudes toward science, mathematics, and engineering fields continues to be a concern. A national study, examining trends in undergraduate education, reveals a steady decline in student interest in the physical sciences and mathematics over the last thirty years (Astin, 1997). Female, African-American, and Hispanic students appear to have a lower level of interest in the sciences than male, Asian, and Caucasian students (National Science Board, 2002). Therefore, the challenge today is two-fold: We must successfully prepare students for careers in science and mathematics, and we must increase students' interest and attitudes toward science, mathematics, engineering, and technology, especially students from diverse backgrounds.

The robotics program examined in this study was designed to increase student attitudes toward science, technology, engineering, and mathematics (STEM) and to increase interest in pursuing STEM-related fields. The robotics program is called the FIRST (For Inspiration and Recognition of Science and Technology) Robotics Competition. FIRST is designed to inspire students in science, mathematics, and technology by combining engineering and technology into a real-world, project-based learning experience (Kamen, 2006). The key component of the robotics program is a six-week design period where teams of high school students design and build a robot designed for a specific task. Teams of 12–20 students worked with a high school teacher and mentors from local universities, professional organizations, and/or businesses to build a robot they could use to compete against other teams from across the United States.

In this study, a sample of high school students who participated in the FIRST Robotics Competition was compared to a sample of students from the same schools who did not participate in the program. Students' attitudes were measured both before and after the six-week design and build period in order to examine the impact of the FIRST Robotics Competition on students' attitudes toward science. The study was designed to answer the following research question:

Do high school students who participate in the FIRST Robotics Competition have a more positive attitude toward science with respect to the perception of scientists?

BACKGROUND LITERATURE

FIRST is a non-profit multinational organization located in Manchester, New Hampshire. The driving goal of FIRST is to make science, mathematics, engineering, and technology as "cool for kids as sports are today" (FIRST, 2006a, np). FIRST has several programs ranging from the Lego League for elementary students, the FIRST Tech Challenge for junior high and high school students, and the FIRST Robotics Competition designed exclusively for high school students ages 14 and above. The FIRST Robotics Competition, which was examined in this study, is designed to build awareness and interest in science and engineering in high school students and to provide them with challenging and engaging learning opportunities (FIRST, 2006a).

The goal of FIRST is to create a world in which young people "dream of becoming science and technology heroes" (FIRST, 2006a, np). When FIRST was founded more than a decade ago, it was modeled to emulate professional sports in America. The goal was to inspire young people to pursue careers in science and technology in the same way professional sports inspires young people to pursue careers as professional athletes (FIRST, 2006a). FIRST uses marketing and media techniques to motivate students to want to learn about science and technology. The FIRST Robotics Competition was patterned after MIT Professor Woodie Flowers' engineering design course. Flowers, who acts as a national advisor to FIRST, believes the key to the competition is that "it celebrates the efforts that came before the actual competition, as well as the gracious professionalism displayed at the competition; and the kids know that we still accept them even if their robots don't work" (Bowden, 1998, p. 19).

Each team, typically consisting of 10–25 students plus technical mentors and faculty advisors, has just six weeks to design and build a robot for the given task out of a common set of basic parts. They must follow rules limiting the size, weight, and cost of their robots. Typically, the robots weigh 100–120 pounds. The goal of the robot, or the "game" as it is described, changes each year and in not announced until the worldwide "Kick-Off" event the first Saturday in January (FIRST, 2006b).

Following the six-week build period, the robots are shipped to regional events, usually held at university arenas. The regional events involve between 40 to 70 teams and last two and a half days. The world championship is held in late April in Atlanta (First, 2006b).

There are empirical studies that show that project-based learning environments, like FIRST, can have a positive impact on student achievement in science and mathematics. A three-year study conducted in British secondary schools found significant differences in both student understanding and academic achievement in mathematics based on standardized test scores as a result of their participation in project-based schools (Boaler, 1999). The study found that three times as many students enrolled in the project-based schools earned the highest possible grade on the national examination in mathematics (Boaler, 1999). A similar study found that after project-based learning activities were used, eighth graders in Union City, New Jersey, scored twenty-seven percentage points higher than students from other urban school districts on statewide tests in reading, math, and writing achievement (Honey & Henriquez, 1996). This is significant given that four years prior to the implementation of project-based learning activities, the state had considered a takeover of the school because it had failed forty of fifty-two indicators of school effectiveness (Honey & Henriquez, 1996).

Theoretical Background

Through the use of hands-on, real-life, problem-solving challenges, FIRST embodies the ideals of constructivism and project-based learning. Theoretically, the program is based on constructivist learning as conceptualized by Jean Piaget and social views of learning as conceptualized by Lev Vygotsky. According to Piaget, students possess an innate need to understand how the world operates and to find order, structure, and predictability in their existence (Eggen & Kauchak, 2001; Piaget, 1952, 1959). According to Piaget, students are motivated by a need to understand the world and use adaptive schemes of assimilation and accommodation to organize knowledge into schemes. Experience with the physical world is critical to the formation of schemes and is found in most classrooms in the form of "hands-on" activities (Ball, 1992; Hartnett & Gelman, 1998). Piaget also emphasized the role of social experience in the learner (DeVries, 1997). It is critical that the learners be allowed to test their findings against those of others. This serves as a balancing effect and motivates the learners to adapt new schemes and compare views with those of others (DeVries, 1997; Eggen & Kauchak, 2001).

While Piaget examined the impact of experience, Vygotsky theorized that participation in social activities was vital to learning (Bredo, 1997; Eggen & Kauchak, 2001; Vygotsky, 1978). Vygotsky believed that learning occurs when students gain specific understanding, and development progresses when this understanding is incorporated into a larger, more complex social context (Vygotsky, 1978). FIRST is designed to capitalize on the constructivist and social aspects of learning by providing students a real-world problem they must solve as a team.

Dethlefs (2002) found that the constructivist learning environment dimensions of Personal Relevance, Shared Control, and Student Negotiation were positively related to student attitudes. Empirical studies provide evidence that constructivist learning environments in science and mathematics can have a positive impact on students' attitudes in both science and mathematics. In a study conducted with sixty-two high school students, Nichols and Miller (1994) found that those students assigned to constructivist learning groups showed that greater gains were made in achievement, efficacy, valuing of algebra, and learning goal orientation. Shymansky, Hedges, and Woodworth (1990) confirmed earlier meta-analysis studies that supported findings that student performance was increased through the use of inquiry-based science curricula dating back to the 1960s.

Student Attitudes in Science

Constructivism and project-based learning are also an important factor in student attitudes and motivation. Motivation is considered one of the most significant determinants of students' success or failure in the classroom (Hidi & Harackiewicz, 2000; Reeve, 1996; Ryan & Connell, 1989). Studies have shown that active involvement in learning activities is more motivating than passive involvement (Zahorik, 1996). In addition, student control and responsibility are also associated with increased motivation, which translates into increased learning and retention of information (Eggen & Kauchak, 2001; Lepper & Hodell, 1989).

In the 1990s, studies were conducted using select groups of students, such as at-risk, urban, or those with various disabilities. One such study highlighted the difficulties of engaging African American students who live in relative poverty when they lack motivation to learn and attend class sporadically (Tobin, Seiler, & Walls, 1999). Motivation was also the subject of a study by Dicintio and Gee (1999). They found that their test group of at-risk students was "unmotivated to learn, defiant in learning situations, and evidence[d] a negative attitude toward school" (p. 234). They concluded that "educators and researchers working with at-risk students should be encouraged to try current, innovative, cognitively based methods of motivating students ... [and that] at-risk students need to learn the skills of self-determination and adaptive motivation in school learning—characteristics that cannot be imparted through motivational practices that control and coerce students" (Dicintio & Gee, 1999, p. 235).

The concept and definition of attitude has been extensively studied (Eiser, 1984; Lemon, 1973; Mueller, 1986; Thurstone & Chavez, 1929). Attitude has been described as a non-observable psychological entity that can only be deduced from a manifested behavior (Adolphe, 2002; Mueller, 1986). Thurstone initially described attitude as "the sum total of a man's inclination and feelings, prejudices and bias, preconceived notions, ideas, fears, threats, and conviction about any specified topic" (Thurstone, 1928, p. 531). He later modified his definition, stating that attitude was the "effect for or against a psychological object" (Thurstone, 1931, p. 261). He recanted this definition in a 1946 commentary, stating he actually believed that attitude was more accurately described in his earlier work as "the intensity of positive or negative effect for or against a psychological object" (Thurstone, 1946, p. 39). Additional definitions of attitude include "a mental or neural state of readiness" (Allport & Hartman, 1935, p. 810), a "consistency in response to social objects" (Campbell, 1950, p. 31), and "the covert response evoked by a value" (Linton, 1945, pp. 111–112).

The term *attitude* encompasses a wide variety of affective behavioral verbs, such as *prefer, accept, appreciate*, and *commit.* In most studies, the term "attitudes" is used to refer to the intrinsic values or interests of the students toward science and mathematics (Dethlefs, 2002). In 2000, Dethlefs conducted a study on the relationship of constructivist learning to students' attitudes and achievement in high school science and mathematics. His findings showed the following results:

- Constructivist learning environments are positively associated with student attitudes in high school biology and algebra.
- Deeper cognitive processing strategies were present when students were allowed to exercise more control in their learning activities.
- Students' enrollment in future elective classes was predicted as a result of their attitudes.
- There is a strong relationship between cooperative group-work and students' interest in school.

Studies have examined the relationship of attitude and achievement in the sciences. Student attitude toward science has been shown to correlate with achievement in the science classroom (Germann, 1988; Napier & Riley, 1985). In 1986, Schibeci & Riley studied the relationship between students' background, perceptions, attitudes, and achievement. Their study showed that gender is related to attitudes and achievement, with females scoring lower on both. Hill, Pettus, and Hedin (1990) found a lack of interest in science careers and lack of participation in science-related activities outside of school among middle and high school girls.

Views of Scientists

There is a large body of literature surrounding the perceptions students have of scientists. Much of the research in this area has used students' drawings to discern what their perceptions are. Findings from these studies have often shown a stereotypical perception of scientists. Strong evidence exists that these perceptions cross all grade levels, gender, ethnic groups, and national boundaries (Finson, 2002).

Mead and Metraux's seminal study in 1957, in which 35,000 high school wrote essays describing their images of a scientists, revealed what has now become the classical stereotypical image of a scientist as one who is an older male in a white lab coat, wearing glasses and working in a laboratory. This image was described by students as both positive and negative, with one student stating, "He is a dedicated man who works not for money or fame or self-glory, but . . . for the benefit of mankind and the welfare of his country," while another stated, "He neglects his family—pays no attention to his wife, never plays with his children" (p. 387). As a result of their research, Mead and Metraux called for major changes in secondary science education, including emphasis on group projects, using terms such as biologist and physicist, rather than scientist, and a shift in teaching and counseling practices to encourage more girls to enter careers in science.

Beardslee and O'Dowd further explored students' image of scientists. They developed a questionnaire that was given to 1,200 undergraduate men and women in four colleges in the northeastern United States. The data showed the scientist was perceived as "a highly intelligent individual devoted to his studies and research at the expense of interest in art, friends, and even family" (998). This image closely mirrored that held by the high school students studied by Mead and Metraux.

Chambers (1983) developed the Draw-a- Scientist-Test (DAST), which had students draw a scientist on a blank sheet of paper. Chambers identified seven specific attributes that consistently appeared in students' drawings: lab coat, eyeglasses, facial hair, symbols of research, and symbols of knowledge, technology, and relevant captions such as formulas. He also identified objects such as light bulbs, basements, and mythical images (i.e., Frankenstein) as having significant meaning. Chambers reports that only 28 of the 4,807 drawings analyzed depicted images of female scientists.

Schibeci (1986) suggests that the media, primarily television, significantly reinforced the stereotypical image of the scientist. According to Schibeci, the modern media portrayed the scientist as a white male who was amoral, insensitive, and obsessive. Schiberci's findings were supported by the earlier work of Gardner (1980) who had shown that cultural models and icons, such as those found in television, movies, and comic books, could contribute to students' mental schema. The stereotypical view of scientists held by students has also been shown to originate, in part, from their teachers, who unknowingly communicate a biased viewpoint in their classrooms (McDuffie, 2001). Scientists were often shown "as serious, sometimes ominous, people who pursue science as solitary investigators working in an environment devoid of social interactions" (McDuffie, 2001, p. 17).

Studies have shown that students can change their perceptions of scientists as a result of interacting with scientists. Bodzin and Gehringer (2001) investigated the impact of a classroom visit from a scientist on schoolchildren. Using fourth and fifth grade students, they asked them to "draw a picture of a scientist doing science" (p. 36). The classes were then visited by scientists and the students were asked to redraw their pictures of a scientist four weeks after the visit. Using the DAST-C checklist, the researchers reported a statistically significant difference between the two visits. The previsit drawings showed "Albert Einstein-like" images and other stereotypical features such as lab coats, eye glasses, and predominately males. The postvisit drawings revealed decreased stereotypical features and more females. Overall, previous research on students' view of scientists indicates that students all too often have stereotypical views of scientists, but those views can be altered through specific instructional activities.

METHODS

Study Participants

This study was conducted in a large U.S. Midwestern metropolitan area. Schools included in this study were located in suburban, urban, and rural areas and included both public and private affiliations. The students in this study were from nine different high schools. All of the students from the FIRST Robotics team at each high school were invited to participate in the study. The students who agreed to participate in the study comprised the treatment group. In an effort to compare students who were as similar as possible, a comparison sample of students was comprised of students from the same schools and from the same science classes as the students on the robotics teams. The pre-survey was administered prior to the six-week ro-

botics build season. The post-survey was administered immediately following the conclusion of the six-week robotics build season.

A total of 132 students completed the pre-survey. Eighty students reported that they were members of robotics teams; fifty-two reported they were
not members of a robotics team. Ninety-nine students who completed the
pre-survey also completed the post-survey. Fifty-eight reported they were
members of robotics teams; forty-one reported they were not members of
a robotics team. Students completing the post-survey represented a 74.43%
participation rate. The treatment group (robotics team members) consisted of 80 participants in the pre-survey and 58 in the post-survey. The
comparison group (non-robotics team members) consisted of 52 in the
pre-survey and 41 in the post-survey. The gender of the students completing the pre-survey was calculated at 36.3% female and 63.8% male in the
robotics team members group and 51.9% female and 48.1% male in the
non-robotics team members group. The gender of the students completing the post-survey was calculated at 37.3% female and 62.7% male in the
robotics team members group and 52.5% female and 47.5% male in the
non-robotics team members group. The race profile of robotics team members completing the post-survey was calculated at 81.4% white and 18.6%
non-white. The race profile of non-robotics team members completing the
post-survey was calculated at 62.5% white and 37.5% non-white.

Inferential statistics were run on the post-survey results to compare the two
groups in order to determine the extent to which the two groups were comparable. The results of a Chi-square test were not significant for females, χ^2
$(1, N = 56) = .07, p < .01$, but were significant for males, χ^2 $(1, N = 76) = 8.90$,
$p < .01$. These results show that the two groups were statistically similar in
regard to the number of female students in the sample, but were not in regard to the number of male students. The results of a Chi-square test were
significant for white students, χ^2 $(1, N = 94) = 13.79, p < .01$. However, they
were not significant for non-white students, χ^2 $(1, N = 38) = 1.68, p < .01$.
These results show that the two groups were statistically similar in regard to
the number of non-white students, but were not similar in regard to white
students.

An independent-samples t test was conducted on FIRST and non-FIRST
groups' overall GPA in science classes to evaluate whether their mean was
significantly different. The test was significant, $t(113) = -2.02, p = .99$. The
students participating in FIRST ($M = 2.09, SD = .67$) reported slightly lower
GPAs in science classes than students not participating in FIRST ($M = 2.45$,
$SD = 1.21$).

An independent-samples t test was conducted on FIRST and non-FIRST
groups' overall GPA to evaluate whether their mean was significantly different. The test was not significant, $t(118) = .182, p = .12$. The students participating in FIRST ($M = 2.44, SD = .86$) reported approximately the same GPA

as students not participating in FIRST (M = 2.41, *SD* = .81). Therefore, the two groups appear statistically similar.

Overall, the results indicate that the two groups appear comparable. There were only two differences between the two groups. The results indicate that the treatment group had slightly more male students but had a slightly lower GPA in science classes.

Instrumentation

The Test of Science-Related Attitude (TOSRA) was used to assess students' attitudes toward science (Fraser, 1981). The TOSRA is designed to measure seven distinct science-related attitudes among secondary school students: Social Implications of Science, Normality of Scientists, Attitude toward Scientific Inquiry, Adoption of Scientific Attitudes, Enjoyment of Science Lessons, Leisure Interest in Science, and Career Interest in Science (Fraser, 1981). The TOSRA is designed to be used by educators and researchers to monitor student progress towards achieving attitudinal aims (Fraser, 1981). In his handbook for TORSA, Fraser states, "TORSA is likely to be most useful for examining the performance of groups or classes of students" (Frasier, 1981, p. 1).

Each scale on TOSRA contains ten items, while the total instrument contains 70 items. A sample of questions from the TOSRA is listed in the Appendix. The response scale is a five-point Likert scale with responses ranging from Strongly Agree (1) to Strongly Disagree (5). Within each scale, five are positive items, five are negative, with respect to their position on science and science-related issues. All items were adjusted so that the higher numeric values associated with the response categories of the items always reflected the positive side of the Likert scale indicating a more positive view of science. The coefficient alpha of .97 suggests that the TOSRA was reasonably reliable for respondents in this study.

Research Questions

This study measured the impact of participation in the FIRST Robotics Competition on students' attitudes toward science, including the interest and attitude subscales on TOSRA. Specifically, this study focused on answering the following research question: Do high school students who participate in the FIRST Robotics Competition have a more positive attitude toward science with respect to the perception of scientists? It was hypothesized that students participating in high school robotic competitions would have a larger change in attitudes and interests in science than students who did not participate in the robotic competition.

RESULTS

Normality of Scientists

A comparison of the pre- and post-means indicated that students who participated in the FIRST Robotics Competition had a more positive attitude toward the normality of scientists than students who did not participate in FIRST, as shown in Table 10.1. Examples of the type of questions on this scale included, "Scientists are about as fit and healthy as other people," and "Scientists do not have enough time to spend with their families."

Students in the FIRST group had statistically significantly higher attitude means ($p = 0.011$) than students in the comparison group regarding their attitude about the normality of scientists as measured by the TOSRA. The strength of relationship between the FIRST factor and the dependent variable was assessed by a partial $\eta^2 = 0.065$, with the FIRST factor accounting for 6.5% of the variance of the dependent variable.

This indicates that students participating in FIRST Robotics have a greater appreciation and more positive attitude towards scientists and members of the scientific community in general than students not participating in FIRST Robotics. During the same time, the non-FIRST students recorded a decrease in the mean. The increase in the mean of the FIRST students may be a result of their interaction with technical mentors and engineers.

A focus group interview was also conducted with a sample of students to gather more details on views of scientists. When students were asked if working with the technical mentors had changed their ideas about scientists and engineers, students spoke excitedly about their relationship with the mentors. One student responded, "They weren't like real geeky or anything. So it shows that not all scientists or engineers are geeky, you know, they can be pretty cool people." A agreed by stating that her mentor was "fun . . . wasn't geeky or nerdy like you would think a lot of guys like him who do this stuff for a living. He was a pretty cool dude." Another student added that "robot-

TABLE 10.1   Normality of Scientists Results for FIRST and non-FIRST Students

Students	Pre-M	SD	Post-M	SD	Difference
FIRST ($N = 59$)	36.18	5.06	37.20	6.29	1.02
Non-FIRST ($N = 40$)	34.60	4.56	33.38	5.06	−1.22

ics has given me a lot more appreciation for computer programming and engineering in general and for the complexities involved there."

DISCUSSION

The more favorable attitudes of FIRST students towards the normality of scientists may be the result of several factors. While participating in the six-week build period, students work directly at designing, building, and testing their team's robot. For many team members, this is the first time outside of a classroom setting that they have had the opportunity to experiment and actually apply skills learned in the classroom. It is during this time that science becomes "real" and not just something found in textbooks.

The results of this study indicate that participants in FIRST view scientist as more "normal" than students who do not participate in FIRST, yet the FIRST students' view of normal includes descriptions such as "geek." This indicates not only a shift in viewpoint toward scientists, but also a shift in the vernacular among FIRST participants. While the general public often uses "geek" in a pejorative context, FIRST students have adopted the term "geek" to represent a position of status and honor, of which they strive to be a part.

Some of the explanation can also be found in the working relationship that is developed during the build season with the technical mentor, many of whom are engineers, who volunteer to work with the teams. For many students, this is their first interaction with scientists and engineers and the close relationship that is built influences their perceptions of the profession. Many students who participate in FIRST discover that the terms "geek" and "nerd" do not apply to these individuals; they perceive them as "cool" and "hip."

The Oxford English Dictionary (OED) defines "geek" as "a simpleton, a dupe, a person who is socially inept or boringly conventional or studious." The Urban Dictionary agrees with the OED definition of "geek" but also adds that in modern usage, the term "enjoys special status within the technical community," and that it "indicates a recognition that most people still consider programming computers to be a bizarre act, along with a certain fierce satisfaction in being very good at their inglorious profession." The term "geek" often carries a positive connotation when used by a member of the group.

Perhaps the explanation for the increase in attitude by students participating in the FIRST Robotics Competition relates to the notion that the competition is built around the foundation of experimentation and inquiry. The focus of the FIRST Robotics Competition is the six-week build season in which teams complete the task of designing, building, testing, and shipping their robots. The six-week build season begins with the "Kick-Off." On the first Saturday of January, the "challenge" is presented via a simultaneous NASA broadcast from FIRST Headquarters in New Hampshire. It is during this broadcast that students find out what the challenge is and specifics about the rules, and they are introduced to that year's kit of parts (the basic building components all teams start with for the competition). Following the broadcasts, teams meet to begin the brainstorming process and to gather the necessary information to best approach their solution to the challenge. The following weeks are filled with design meetings, building prototypes, construction of the robot, testing, redesigning, and finally shipment of the finished robot to a regional event. It is during these six weeks that the students work as much as possible, often well into the night and weekends. This is when the students actually apply the science and mathematics they learn in the classroom and when they work most closely with the technical mentors.

While the previous studies on the impact of the FIRST Robotics Competition did not use the TOSRA, they do show similar outcomes as a result of student participation in the program. In 1998, Atlantic Associates reported on students' attitudes and skills as affected by FIRST Robotics. Their survey included not only the students participating in FIRST, but also included the parents of the students, school personnel, and corporate partners involved with the teams. The results showed that all vested parties in the program saw positive impact on students' problem-solving ability, teamwork skills, self-confidence, and attitudes toward careers in engineering (Atlantic Associates, 1998).

The Goodman Research Group conducted a survey of the FIRST teams during the 2000 FIRST Robotics Competition season. According to the results of the survey, FIRST attracts boys and girls of all different grade levels with each group having different levels of interest and commitment to science, technology, engineering, and mathematics. Pre- and post- survey analysis showed statistically significant increases in participants' attitudes toward teamwork and positive self-image (Goodman Research Group, 2000).

All in all, this study on students' attitudes toward scientists has important implications for the field of science education. The results imply that programs that engage students in authentic scientific problems can significantly improve students' attitudes and views of science. The FIRST program appears to have helped students develop a more positive view of scientists. Positive views toward science are often viewed as an important correlate to achievement in science, and as a result, programs like FIRST can be an important part of helping students achieve in science. The positive attitudes and interest may also lead to future careers in science related fields, but more research is needed to better understand the long-term impact of the program like FIRST on students.

REFERENCES

Adolphe, F. (2002). *A cross-national study of classroom environment and attitudes among junior secondary science students in Australia and in Indonesia.* Doctoral dissertation, Curtin University of Technology. Retrieved March 22, 2006 from http://adt.curtin.edu.au/theses/available/adt-WCU20031201.141540

Allport, F.H. & Hartman, D.A. (1935). Measurement and motivation of a typical opinion in a certain group. *American Political Science Review, 19,* 735–760.

Astin, A. (1997). How "good" is your institution's retention rate? *Research in Higher Education, 38*(6), 647–658.

Atlantic Associates. (1998). *1998 competition survey results.* Manchester, NH: Author.

Ball, D. (1992). Magical hopes: Manipulatives and the reform of mathematics education. *American Educator,* Summer, 28–33.

Beardslee, D.C., & O'Dowd, D.D. (1961). The college-student image of the scientist. *Science, 122*(3457), 997–1001.

Boaler, J. (1999). Mathematics for the moment, or the millennium? *Education Week.*

Bodzin, A., & Gehringer, M. (2001). Breaking science stereotypes. *Science and Children, 38*(4), 36–42.

Bowden, T. (1998). Robotics: The ultimate mind sport. *Tech Directions, 57*(10), 18–20.

Bredo, E. (1997). The social construction of learning. In G. Phye (Ed.), *Handbook of academic learning: Construction of knowledge* (pp. 3–45). San Diego: Academic Press.

Brown, L. (Ed.). (1993). *The New Shorter Oxford English Dictionary.* Oxford: Clarendon Press.

Campbell, D.T. (1950). The indirect assessment of social attitudes. *Psychological Bulletin, 47,* 15–38.

Chambers, D.W. (1983). Stereotypic images of the scientist: The Draw-a-Scientist Test. *Science Education, 67*(2), 255–265.

Dethlefs, T. M. (2002). Relationship of constructivist learning environment to student attitudes and achievement in high school mathematics and science. *Dissertation Abstracts International, 63*(07), 2455.

DeVries, R. (1997). Piaget's social theory. *Educational Researcher, 26*(2), 4–18.

Dicintio, M. J. & Gee, S. (1999). Control is the key: Unlocking the motivation of at-risk students. *Psychology in the Schools, 36*(3), 231–237.

Eggen, P. D., & Kauchak, D. (2001). *Educational psychology: Windows on classrooms.* Upper Saddle River, NJ: Merrill Prentice Hall.

Eiser, J.R. (Ed.). (1984). *Attitudinal judgment.* New York: Springer-Verlag.

Finson, K.D. (2002). Drawing a scientist: What we do and do not know after fifty years of drawing. *School Science and Mathematics, 102*(7), 335–345.

FIRST. (2006a). *About FIRST.* Retrieved from http://www.usfirst.org/about

FIRST. (2006b). *FIRST Robotics Team Information.* Retrieved from http://www.usfirst.org/teaminfo

Fraser, B. J. (1981). *TOSRA: Test of science-related attitudes handbook.* Hawthorn, Victoria: Australian Council for Educational Research.

Gardner, P.L. (1980). The identification of specific difficulties with logical connectives in science among secondary school students. *Journal of Research in Science Teaching, 17*(3), 223–239.

Germann, P. J. (1988). Development of the attitude toward science in school assessment and its use to investigate the relationship between science achievement and attitude toward science in school. *Journal of Research in Science Teaching, 25,* 689–703.

Goodman Research Group. (2000, November). *Final report to FIRST.* Cambridge, MA.

Hartnett, P. & Gelman, R. (1998). Early understanding of numbers: Paths or barriers to the construction of new understandings. *Learning and Instruction, 8*(4), 341–374.

Hidi, S., & Harackiewicz, J. M. (2000). Motivating the academically unmotivated: A critical issue for the 21st century. *Review of Educational Research, 70,* 151–179.

Hill, O. W., Pettus, W. C., & Hedin, B. A. (1990). Three studies of factors affecting the attitudes of blacks and females toward the pursuit of science and science related careers. *Journal of Research in Science Teaching, 27,* 289–314.

Honey, M. & Henriquez, A. (1996). *Union City Interactive Multimedia Education Trial: 1993-1996 Summary Report.* Education Development Center, Center for Children and Technology. CCT Reports Issue No. 3.

Kamen, D. (2006, June 20). Phone interview.

Lemon, N. (1973). *Attitudes and their measurement.* London: Batsford.

Lepper, M. & Hoddell, M. (1989). Intrinsic motivation in the classroom. In C. Ames & R. Ames (Eds.); *Research on motivation in education* (Vol. 3 pp. 73–105). San Diego: Academic Press.

Linton, R. (1945). *The cultural background of personality.* New York: Appleton-Century-Crofts.

McDuffie, T.E. (2001). Scientists—Geeks & nerds? *Science and Children, 38*(8), 16–19.

Mead, M., & Metraux, R. (1957). Image of the scientist among high school students: A pilot study. *Science, 126,* 384–390.

Mueller, D.J. (1986). *Measuring social attitudes.* New York: Teachers College Press, Columbia University.

Napier, J. D., & Riley, J. P. (1985). Relationship between affective determinants and achievement in science for seventeen year olds. *Journal of Research in Science Teaching, 22,* 365–383.

National Science Board. (2002). *Science and engineering indicators 2002.* Two Volumes (volume 1, NSB-02-1; volume 2, NSB-02-1A). Arlington, VA: National Science Foundation.

National Science Foundation. (2002). *S&E labor market conditions.* Retrieved from http://www.nsf.gov

Nichols, J. E., & Miller, R. B. (1994). Cooperative learning and student motivation. *Contemporary Educational Psychology, 19,* 167–178.

Ortega, S. (2003). Projects, process and pipelines: Challenges to enhancing the scientific labor force. Reprinted from the white paper: Rice, M.L. (Rd.). (2003). *Recruiting and training future scientists: How policy shapes the mission of graduate education* (MASC Report No. 107). Lawrence, KS: University of Kansas Merrill Advanced Studies Center.

Peckham, A. (Ed.). (2005). *Urban Dictionary*. Kansas City: Andrews McMeel Publishing.

Piaget, J. (1952). *Origins of intelligence in children*. New York: Humanities Press.

Piaget, J. (1959). *The language and thought of the child* (3rd ed.) (Gabain, M. & Gabain, R. trans.). New York: Routledge/Taylor & Francis Group.

Reeve, J. (1996). *Motivating others*. Boston: Allyn & Bacon.

Ryan, R. M., & Connell, J. P. (1989). Perceived locus of causality and internalization: Examining reasons for acting in two domains. *Journal of Personality and Social Psychology, 57*, 749–761.

Schibeci, R. A., & Riley II, J. P. (1986). Influence of students' background and perceptions on science attitudes and achievement. *Journal of Research in Science Teaching, 23*, 177–187.

Shymansky, J. A., Hedges, L. V., & Woodworth, G. (1990). A reassessment of the effects of inquiry-based science curricula of the 60's on student performance. *Journal of Research in Science Teaching, 27*, 127–144.

Thurstone, L.L. (1928). Attitudes can be measured. *American Journal of Sociology, 38*, 268–389.

Thurstone, L.L. (1931). The measurement of social attitudes. *Journal of Abnormal and Social Psychology, 26*, 249–269.

Thurstone, L.L. (1946). Comment. *American Journal of Sociology, 52*, 39–50.

Thurstone, L.L., & Chavez, E. J. (1929). *The measurement of attitude*. Chicago: University of Chicago Press.

Tobin, K., G. Seiler, & Walls, E. (1999). Reproduction of social class in the teaching and learning of science in urban high schools. *Research in Science Education, 29*(2), 171–187.

Vygotsky, L. S. (1978). *Mind in society: The development of higher psychological processes*. Cambridge, MA: Harvard University Press.

Wilkinson, R. K. (2002). How large is the U.S. S&E workforce? *InfoBrief*, NSF-02-325. Retrieved April 4, 2004, from http://www.nsf.gov/sbe/srs/infbrief/nsf0

Zahorik, J. (1996). Elementary and secondary teachers' reports of how they make learning interesting. *The Elementary School Journal, 96*(5), 551–564.

APPENDIX

Sample Questions Normality of Scientists

Q2 "Scientists usually like to go to their laboratories when they have a day off."

Q9 "Scientists are about as fit and healthy as other people."

Q16 "Scientists do not have enough time to spend with their families."

Q23 "Scientists like sports as much as other people do."

Q30 "Scientists are less friendly than other people."

Q37 "Scientists have a normal family life."

Q44 "Scientists do not care about their working conditions."

Q51 "Scientists are just as interested in art and music as other people."

Q58 "Few scientists are happily married."

Q65 "If you met a scientist, he would probably look like anyone else you might meet."

CHAPTER 11

ATTITUDES TOWARDS SCIENCE AND SCIENTIFIC METHODOLOGY WITHIN A SPECIFIC PROFESSIONAL CULTURE

Darko Hren
University of Split, Croatia

Modern medicine largely relies on scientific work and evidence for best practice. Evidence Based Medicine (EBM) is a relatively new term—it was first used in the 1990s (Claridge & Fabian, 2005), but the scientific way of thinking and understanding in medicine can be traced, in its rudimentary form, as far back as classical Greece (Claridge & Fabian, 2005). Sackett, Rosenberg, Gray, Haynes, and Richardson (1996) defined the practice of EBM as integrating individual clinical expertise with the best available external clinical evidence from systematic research. This means that an integral and important part of a clinician's everyday work includes using scientific information. The need for understanding and critically appraising this information is today greater than ever before since the number of scientific

Attitude Research in Science Education, pages 281–290

articles published per year has increased from 700,000 in 1990 to 1.4 million in 2008 (Hammond, 2010). Scientific production has definitely risen, but there is a lack of physicians who engage in research (Tugwell, 2004) which is, along with clinical care and teaching, one of three traditional pillars of academic medicine (Marušić, 2004).

Although the relationship between attitudes and behavior has been a subject for debate since the earliest days of social psychology (Shultz & Oskamp, 2005), fostering students' attitudes towards science can be viewed as a desirable outcome from a perspective of preparing them to engage in EBM approach.

However, attitudes are not affected easily (Rajecki, 1990), and even if the course on scientific methodology did affect students' attitudes towards science, that effect cannot be expected to be large or permanent. It can be an important initial *push*, but the broader climate among their teachers and other physicians must exist because it provides a strong and stable influence, through formation of students' social identity (Tajfel & Turner, 1986) as members of the medical professionals group. Therefore, it is important that other medical professionals also regard science highly positively. This especially applies to medical teachers who serve as role models to many generations of future physicians.

This chapter will present a line of research performed in medical setting aiming to establish medical students' attitudes towards science and the effect of a course on principles of scientific research in medicine on those attitudes. Additionally, medical teachers' and general practitioners' attitudes towards science were investigated as indicators of a broader professional climate.

TEACHING PRINCIPLES OF SCIENTIFIC RESEARCH TO MEDICAL STUDENTS

In 1995, Zagreb University School of Medicine introduced a graduate course called Principles of Scientific Research in Medicine into the curriculum. It was a pioneering endeavor among Croatian medical schools, but also in the international context because principles of scientific research are rarely taught in a comprehensive way at medical schools (Parkes, Hyde, Deeks, & Milne, 2002). The course was set in the second year of the medical curriculum with the idea of providing students with introductory EBM principles and knowledge of how to critically appraise scientific information before they enter clinical rounds. Its scope focused on five themes: (1) principles of scientific research, (2) accessing bibliographic databases and finding information, (3) practical approaches to the collection and presentation of data, (4) principles of assessing and writing a scientific article, and (5) responsible conduct of research (Marušić & Marušić, 2003).

As the course took roots, teachers who taught it decided to assess its effects. Knowledge, of course, is the most obvious outcome, but one of the aims—which might even be called a *hidden* aspect of the course's curriculum—was fostering students' positive attitudes towards science and scientific thinking (Lemp & Seal, 2004).

ZAGREB SCIENCE ATTITUDE SURVEY

As there was no adequate instrument that could be used to measure attitudes towards science in medical setting, researchers developed a new one. It was a 45-item Likert type scale, and it served as the basis for all the projects along this research line (Cvek et al., 2009; Hren et al., 2004; Mrdeša-Rogulj et al., 2007; Vujaklija et al., 2009). A Likert type scale was chosen because its construction is relatively simple and interpretation of results straightforward (Anastasi, 1988). The instrument was intended to be a unidimensional measure of attitudes towards science in a medical setting, but during its development three subscales could be distinguished—"Value of science to medicine," "Value of science to humanity," and "Value of scientific methodology" (Table 11.1). Reliability of the instrument was high in all examined samples. Cronbach alpha was above 0.9 for the whole scale, and above 0.7 for single subscales (Cvek et al., 2009; Hren et al. 2004; Mrdeša Rogulj et al., 2007; Vujaklija et al., 2009). A more detailed view of the development of the questionnaire can be found elsewhere (Hren et al, 2004).

TABLE 11.1 Zagreb Science Attitude Survey

Value of Science for Medicine:
1. Valid medical procedures are only those that have been verified by clinical research.
2. Scientific methodology allows us to establish the exact cause of disease exactly.
3. Science made the greatest contribution for development of medicine.
4. Scientific methodology misleads the physician.
5. Only the scientific kind of research makes the real progress in medicine possible.
6. Scientific approach is a limitation for the physician.
7. Use of scientific methodology is basic for progress in medicine.
8. By keeping up with science, a physician can keep up with time.
9. Scientific approach in medicine is needless.
10. Science distracts physicians from natural ways of healing.
11. Treatment methods that are not scientifically verified are dangerous and unreliable.
12. Knowledge of scientific methodology does not influence my medical competence.
13. Scientific methodology allows the progress in medicine.
14. Only scientifically verified methods of treatment are reliable.
15. Science has extended human life span.

(continued)

TABLE 11.1 (continued)   Zagreb Science Attitude Survey

16. Science is indispensable if cure for cancer is to be found.
17. New medicines can only be discovered by scientific approach.

Value of Science for Humanity:

1. Science has too rigid view about the world.
2. Development of science has very much improved the quality of life.
3. Science often oversteps the limits of what is morally allowable.
4. Progress of the human society would be impossible without science.
5. Scientific approach is devoid of all humaneness.
6. Scientists often use unethical methods in their research.
7. Scientific approach is rigid and inapplicable in real life situations
8. Science makes the solving of ecological problems easier.
9. Science enables us to better understand the world.
10. If science continues in the same direction it has so far, it will be the end of humanity.
11. Science has caused great catastrophes during the history.
12. Negative effects of science exceed the positive ones.
13. If there were no science, we would lead a healthier and more peaceful life.
14. Scientists unethically use innocent animals for their experiments.
15. There would be no progress without scientific progress.
16. Science is the main cause of the ecological catastrophe that threats us.

Value of Scientific Methodology:

1. Sound discoveries are not possible without scientifically based research.
2. High-quality research can be done only with good knowledge of methodology.
3. Scientific methods impose unnecessary rules.
4. The only data that we can be sure of are those obtained by scientific methods.
5. Scientific way of thinking is dry and boring.
6. Only by scientific approach can one obtain objectively measurable and precise data.
7. Scientific approach allows easier understanding of the problem.
8. Scientific methodology only makes the performance of medical research more difficult.
9. Studying scientific methodology only places an unnecessary burden on students.
10. Every physician must have a good knowledge of scientific methodology.
11. Facts can be established only scientifically.
12. Knowledge of scientific methodology is necessary for obtaining correct and objective data.

ATTITUDES TOWARDS SCIENCE AND SCIENTIFIC METHODOLOGY WITHIN MEDICAL PROFESSIONAL CULTURE

The scale was first applied at the beginning of the academic year 2001/02 (Hren et al., 2004). Students from all six study years were surveyed at the end of the first lecture they attended (N_{total} = 932, which was 58% of all students at Zagreb University School of Medicine). Third-year students were especially interesting because they were the ones who just finished and passed the course Principles of Scientific Research in Medicine, which was taught in the summer semester of the second year.

Overall, their attitudes towards science were positive, with average scores around 70% of the theoretical maximum[1] and significantly differing from a neutral position (i.e., from 50% of theoretical maximum).

Third-year students, who had just finished the course Principles of Scientific Research in Medicine, had the most positive attitudes. The effect sizes were in the medium-size range, Cohen's d ranging from 0.3 to 0.6 (Cohen, 1992), and this was a promising result in terms of potential effects of the course. Students' attitudes were positively associated with their grades and score on a short knowledge test about scientific methodology that was applied along with the survey. They were not associated to students' gender or failure to pass a year.

This cross-sectional study provided an initial implication that attitudes might change during medical school because significantly more positive attitudes were registered in third-year students. The next task was to investigate whether the changes could be registered in repeated measurements, and if so, whether they could be related to teaching students about scientific research in medicine.

Having already surveyed first-year students in the first study, researchers continued to follow this generation throughout the remaining five years of their medical education. A shortened, 18-item version of the survey was applied at the beginning of each subsequent academic year, from 2002/03 to 2006/07 (Vujaklija et al., 2009). The shortened version was used in order to rationalize the time needed for completion of the survey. It included 13 items from the subscale "Value of science for medicine" and five from "Value of science for humanity." Internal consistency of this version was good (between 0.75–0.80 in repeated applications of the instrument), and principal components analysis showed that it could be considered as a unidimensional scale. Questionnaires were collected only from students who indicated that they enrolled the current study year for the first time, in order to prevent obtaining data from the same students twice in the same study year. Although this may be viewed as a potential source of selection bias, a previous study showed no association between attitudes and failure to pass a year (Hren et al., 2004).

This study again showed that students had positive attitudes towards science, but also that those attitudes significantly increased, from 64% of the theoretical maximum in the first year to 73% in the second ($p < 0.001$) and then to 78% in the third year ($p = 0.012$). However, once the effect of students' GPA on their attitudes was statistically excluded, these differences became insignificant ($p = 0.058$). Given the marginal insignificance of the p-value, researchers also performed a polynomial contrasts analysis for trend and found that attitudes scores followed a statistically significant cubic trend ($p = 0.011$) of initial increase up to the third year of the study, followed by a decrease in fourth and fifth, and final increase in the sixth year of the study (Vujaklija et al., 2009).

The final step, and a second part of the study by Vujaklija and colleagues, was a direct investigation of the effects of the course Principles of Scientific Research in Medicine. Organization of classes at the Zagreb University School of Medicine allowed researchers to organize a non-randomized controlled trial with second-year medical students in the academic year 2006/07. Courses were organized in blocks, with students attending one course at a time. However, not all students attended the same course at the same time. They were divided in two groups, so that one group would have, for example, Physiology and Immunology, while the other would have Principles of Scientific Research in Medicine. Later in the year they would change and by the end of the year all courses were given twice.

All second-year medical students were surveyed in November of 2006, before any of them had taken the course on scientific methodology. At that time they were also asked to fill in a six-character code so that researchers could match their scores in repeated measurements. Anonymity was preserved as the codes were not related to any identifiable information about students.

In January 2007, the first group had the course block, and all students were once again surveyed four months later in May 2007, just before the second group had the course. The codes were matched for 64 students who had the course between the two measurements (intervention group) and 80 who had not taken the course (control group). The obtained results showed the effect of the course as the two groups did not differ in the first measurement, but in the second measurement students from the intervention group had significantly more positive attitudes ($p < 0.001$, Cohen's $d = 0.3$; Vujaklija et al., 2009).

In subsequent research projects, medical teachers' (Cvek et al. 2009) and general practitioners' (Mrdesa-Rogulj et al., 2007) attitudes towards science were measured. When 327 medical teachers (177 instructors and research fellows and 150 professors) from Zagreb University School of Medicine were surveyed, their scores averaged above 80% of the theoretical maximum, which were the highest values obtained in all the samples surveyed along this line of research. Moreover, professors had significantly higher attitudes compared to instructors (Cohen's $d = 0.3$), and attitudes towards science were also the strongest predictor of number of published articles,[2] even after controlling for age and rank of medical teachers.

Finally, general practitioners also had positive attitudes towards science, just below 80% of theoretical maximum (Mrdesa-Rogulj et al., 2007). Their scientific production, however, was low—only 6% of 427 surveyed general practitioners have ever published an article in a journal indexed in PubMed.

It seems that the relationship between attitudes towards science and number of published articles was mediated by requirements for scientific production imposed on medical teachers. Once they *had* to do science, those with more positive attitudes did better, but when there were no requirements, like in the case of general practitioners, a large majority did not partake in scientific work. This is a hypothesis that still needs to be investigated and evaluated.

CONCLUDING REMARKS

Our line of research revealed positive attitudes towards science within three groups of medical professionals—medical students, teachers, and general practitioners. One limitation may be that all the findings were obtained on Croatian samples. However, our instrument was used by other researchers on other samples of medical professionals and indicated generally positive attitudes towards science in other settings as well (Burazeri et al., 2005; Khan, Khawaya, Waheed, Rauf, & Fatmi, 2006; Pruskil, Burgwinkel, Georg, Keil, & Kiessling, 2009).

Furthermore, we showed that it is possible to influence students' attitudes early in the medical curriculum, but also that a single course is not enough to obtain a permanent change. This is important information, especially in the context of expansion of the academic research enterprise (Mallon, 2006), which is not paralleled by a corresponding number of physician-investigators (Ley & Rosenberg, 2005). Instead of addressing the issue of raising the interest in scientific research at the MD/PhD level, a more fruitful approach may be systematic work with medical students (Langhammer, Garg, Neubauer, Rosenthal, & Kinzy, 2009; Solomon, Tom, Pichert, Wasserman, & Powers, 2003; Ziel, Friedman, & Smith, 2006).

A vertical integration of a course on scientific methodology into medical schools may provide stronger and more stable effects on students' attitudes and interest in science. It may be especially effective if a practical research experience was required from students, from conception and design of the study to writing, and potentially publishing a research paper.

Fazio (1986) and his colleagues have postulated a model of attitude–behavior relationship which predicts that the accessibility of the attitude moderates the attitude–behavior link. This means that people with favorable but not easily accessible attitudes are less likely to act in ways consistent with their attitudes than those with more accessible attitudes. One way of making positive attitudes towards science more accessible is by providing repeated and regular exposure to scientific reasoning, methodology, and work.

This was the case with medical teachers who had to think about their research and scientific production, and in that group we found a positive association between attitudes and behavior. The same could be done with medical students if engagement in scientific work were required as a part of several linked courses throughout the medical curriculum.

Positive and accessible attitudes towards science could enhance students' application of scientific approach later in their career. It can be through a direct engagement in scientific research or through critical appraisal of the information they come in contact with as practicing physicians.

NOTES

1. Percentages of the theoretical maximum are used as indicators of attitudes for two reasons. First, they allow a straightforward understanding of results than offering raw scores, and second, they allow comparisons among samples investigated in other project with shortened version of the original 45-item questionnaire.
2. Publications were defined as articles published in journals indexed in Current Contents® or Web of Science® databases, required for academic promotion in university medical schools in Croatia.

REFERENCES

Anastasi, A. (1988). *Psychological testing.* New York: MacMillan.

Burazeri, G., Čivljak, M, Ilakovac, V, Janković, S., Majica-Kovačević, T., Nedera, O., Roshi, E., Sava, V., Šimunović, V., Marušić, M., & Marušić, A. (2005). Survey of attitudes and knowledge about science in medical students in southeast Europe. *British Medical Journal, 331,* 195–196.

Claridge, J.A., & Fabian, T. (2005). History and development of evidence-based medicine. *World Journal of Surgery, 29*(5), 547–553.

Cohen, J (1992). A power primer. *Psychological Bulletin, 112,* 155–159.

Cvek, M., Hren, D., Sambunjak, D., Planinc, M., Mačković, M., Marušić, A., & Marušić, M. (2009). Survey of medical teachers' attitudes towards science and motivational orientation for medical research. *Wiener Klinsche Wochenschrift, 121,* 256–261.

Fazio, R. H. (1986). How do attitudes guide behavior? In R. M. Sorrentino & E. T. Higgins (Eds.), *The handbook of motivation and cognition: Foundations of social behavior* (pp. 204–243). New York: Guilford.

Festinger, L. (1957). *A theory of cognitive dissonance.* Oxford, UK: Row, Peterson.

Hammond, K. (2010). Impact factor: Can a scientific retraction change public opinion? *Scientific American,* Retrieved from http://www.scientificamerican.com/article.cfm?id=retraction-impact-lancet&page=3

Hren, D., Lukić, I.K., Marušić, A., Vodopivec, I., Vujaklija, A., Hrabak, M., Marušić, M. (2004). Teaching research methodology in medical schools: students' attitudes towards and knowledge about science. *Medical Education. 38*, 81–86.

Khan, H., Khawaya, M.R., Waheed, A., Rauf, M.A., & Fatmi, Z. (2006). Knowledge and attitudes about health research amongst a group of Pakistani medical students. *BMC Medical Education, 6*, 54.

Langhammer, C.G., Garg, K., Neubauer, J.A., Rosenthal, S., & Kinzy, T.G. (2009). Medical student research exposure via a series of modular research programs. *Journal of Investigative Medicine, 57*, 11–17.

Lemp, H. & Seale, C. (2004). The hidden curriculum in undergraduate medical education: qualitative study of medical students' perceptions of teaching. *British Medical Journal, 329*, 770–773.

Ley, T.J. & Rosenberg, L.E. (2005). The physician-scientist career pipeline in 2005. Build it and they will come. *JAMA, 294*, 1343–1351.

Mallon, W.T. (2006). The benefits and challenges of research centers and institutes in academic medicine: findings from six universities and their medical schools. *Academic Medicine, 81*, 502–512.

Marušić, B. (2004). Academic medicine: one job or three? *Croatian Medical Journal, 45*, 243–244.

Marušić, A., & Marušić, M. (2003). Teaching students how to read and write science: A mandatory course on scientific research and communication in medicine. *Academic Medicine, 78*(12), 1235–1239.

Mrdeša Rogulj, Z., Baloević, E., Đogaš, Z., Kardum, G., Hren, D., Marušić, A., & Marušić, M. (2007). Family medicine practice and research: Survey of physician' attitudes towards scientific research in a postcommunist transition country. *Wiener Klinische Wochenschrift, 119*, 164–169.

Parkes, J., Hyde, C., Deeks, J., & Milne, R. (2001). Teaching critical appraisal skills in health care settings. *Cochrane Database of Systematic Reviews, 3*.

Pruskil, S., Burgwinkel, P., Georg, W., Keil, T., & Kiessling, C. (2009). Medical students' attitudes towards science and involvement in research activities: A comparative study with students from a reformed and a traditional curriculum. *Medical Teacher, 31,* 254–259.

Rajecki, D.W. (1990). *Attitudes.* Sunderland, MA: Sinauer Associates.

Sackett, D, Rosenberg, W., Gray, M., Haynes, B., & Richardson, S. (1996). Evidence based medicine: what it is and what it isn't. *BMJ, 312*, 71–72.

Shultz, P.W., & Oskamp, S. (2005). *Attitudes and opinions.* Mahwah, NJ: Lawrence Erlbaum.

Solomon, S.S., Tom, S.C., Pichert, J., Wasserman, D., & Powers, A.C. (2003). Impact of medical student research in the development of physician-scientists. *Journal of Investigative Medicine, 51*, 149–156.

Tajfel, H., & Turner, J. C. (1986). The social identity theory of inter-group behavior. In S. Worchel & L. W. Austin (Eds.), *Psychology of intergroup relations* (pp. 7–24). Chicago: Nelson-Hall

Tugwell, P. (2004). The campaign to revitalize academic medicine kicks off: We need a deep and broad international debate to begin. *Crotian Medical Journal. 45*(3), 241–242.

Vujaklija, A., Hren, D., Sambunjak, D., Vodopivec, I., Ivaniš, A., Marušić A., & Marušić, M. (2009). Can teaching research methodology influence students' attitude towards science? Cohort study and non-randomized trial in a single medical school. *Journal of Investigative Medicine, 58*(2), 282–286.

Ziel, K., Friedman, E., & Smith, L. (2006). Supportive programs increase medical students' research interest and productivity. *Journal of Investigative Medicine, 54*, 201–207.

CHAPTER 12

ATTITUDE RESEARCH IN SCIENCE EDUCATION

Looking Back, Looking Forward

Myint Swe Khine and Issa M. Saleh
Bahrain Teachers College, University of Bahrain

INTRODUCTION

The research into how students' attitudes affect their learning of science-related subjects has been one of the core areas of interest among science educators. The development in science education records various attempts at measuring attitudes and determining the correlations between behavior, achievements, career aspirations, gender identity, and cultural inclination. Some researchers noted that attitudes can be learned and teachers can encourage students to like science subjects through persuasion. But some hold the view that attitude is situated in context and has much to do with upbringing and environment. The critical role of attitude is well recognized in advancing science education, and particularly designing curriculum and choosing powerful pedagogies and nurturing students.

Since Noll's (1935) seminal work on measuring the scientific attitudes, a steady stream of research papers describing the development and valida-

Attitude Research in Science Education, pages 291–296
Copyright © 2011 by Information Age Publishing
All rights of reproduction in any form reserved.

tion of scales have appeared in scholarly publications. Despite these efforts, the progress in this area has been stagnated by limited understanding of the conception of attitude, dimensionality, and an inability to determine the multitude of variables that made up such concept. This book makes an attempt to take stock and critically examine classical views on science attitudes and explore contemporary attempts in measuring science-related attitudes. The chapters in this book are a reflection of researchers who work tirelessly in promoting science education and highlight the current trends and future scenarios in attitude measurement.

ATTITUDE RESEARCH

The first chapter summarizes the key findings from robust research about attitude. Some of the findings that are included in this chapter come from social psychological research (Eagly & Chaiken, 1993). Reid starts the chapter by pointing out that the area of attitude measurement began receiving considerable attention after World War II (Hovland, Janis, & Kelly, 1953). The problem with the study of attitude is that the scaling techniques that are used are incapable of offering the insights needed to help with the progress. The chapter tries to gather evidence to point out some of the problems with the assumptions behind the methodologies that are used in other studies.

In Chapter 2, the authors look at two new approaches to the study of students' responses to school science. The author uses a mixed methods methodology to complete the study. The sample for the study was taken from a Danish upper secondary school. The study looks at the attitude of students towards physics in various types of Cultural Border Crossings. The study uses the following predictors: particular value-related crossings, physics self-concept, and teacher's interest in students. The second approach the authors use to deal with a systematical investigation is "identity of urban upper secondary students" and students' situated responses.

The aim of Chapter 3 is to develop a questionnaire—investigating teachers' beliefs, attitudes and intentions—that deals with the use of simulations for teaching purposes. The authors point out the lack of instrument that focuses on teachers and their use of simulation when teaching. The authors identify the fact that most of the questionnaires that are available focus on technology or computer technologies and argue that the questionnaires are not proper for "investigating teachers' beliefs, attitudes and intentions towards the use of simulations for teaching purposes." The authors use 500 elementary school teachers to do the study.

In Chapter 4, the authors discuss the criticism of attitude scale in science education research literature (Blalock et al., 2008; Gardner, 1996; Lederman, 2007; Munby, 1997; Osborne, Simon, & Collins, 2003; Reid, 2006; Schibeci, 1984). The authors explain that the reason for this criticism is due to too much concentration on studies developing attitude measures on indicators for internal consistency (Gardner, 1996). The overall aim of the chapter is to argue two points in response to the critiques against using "simple" Likert-based questionnaires for measuring scientific attitudes. The first deals with the validity and reliability of attitude scales. The authors bring evidence from conceptual analysis to show what attitude scale is able to measure. The second deals with the steps and techniques to make good attitude scales. The techniques that have been used, according to the authors, have been impacted by the paradigm shift from classical to modern test theory (Bond & Fox, 2001).

In Chapter 5, Yuqing Yu and Felicia Moore Mensah from Columbia University Teachers College discuss the importance of science educators clearly understanding the relationship between Science-Technology-Society (STS) and decision making about socio-scientific issues. The authors argue in this chapter that the Views on Science-Technology-Society (VOSTS) instrument does not allow the application of quantitative analysis. According to the authors, this has to do with the nature of the instrument—which is qualitative in nature. In order to circumvent this problem, the authors suggest using the Multiple Response Model (MRM). The aim of this chapter involves the application of MRM to the VOSTS instrument.

In Chapter 6, "Tailoring Information to Change Attitudes: A Meta-Structural Approach," See and Khoo describe the importance of structural and meta-attitudinal perspectives in measuring attitudes toward science. According to the authors, affective or cognitive structural bases of attitudes represent associations among emotions and beliefs, while cognitive meta-bases of attitudes refer to subjective judgments driven by emotions or beliefs. They present the results from a study that suggests differences in affective and cognitive structural bases and meta-bases for attitudes towards some science subjects. In conclusion, they describe the implications for designing messages to increase interest in science education.

In Chapter 7, Rebello and her colleague describe assessment practices for understanding science-related attitudes. They note that each assessment method presents its own advantages to provide an accurate and true reflection of one's attitudes to a particular aspect of science. They suggest that while there are many ways to assess the attitudes, an appropriate way to examine the assessment practices is through the lens of COI (cognition, observation and interpretation) assessment triangle. The authors assert that the COI assessment triangle provides a unique approach to analyze current attitude-based assessments and assessment design. By using COI, the team analyzed two attitude measurement instruments. One such study involves a case study investigating students' attitudes toward science in relation to their epistemology and conceptual understanding of biotechnology. The authors conclude that attitudes toward science research often focus on just one of the assessment triangle elements and they suggest that more studies are needed to explore the details of each element and how elements relate each other.

Coll and Paku in Chapter 8 raise the issue of indigenous people being under-represented in higher education in general and in science and engineering in particular. They note that indigenous people felt alienated from the Western idea of science. Again, like other authors in this volume, Coll and Paku recognize science as culture and how scientific thinking can be influenced by culture. The chapter presents a program where students working with scientists and learning about the culture of science by doing science. Work-Integrated Learning (WIL) is a collaborative program designed to combine aspects of study and work. WIL combines students' work experience with on-campus academic learning. The authors feel that a work-integrated learning program can be considered a form of enculturation of students to science. They assert that providing an active learning experience may have positive effects on improving students' perceptions of science self-efficacy, their enjoyment of science and their attitude toward science. The chapter concludes by stating that creating opportunities for both indigenous and non-indigenous students to work with professional scientist mentors can improve the students' attitudes towards science-related skills.

In Chapter 9, the author presents the importance of teaching socio-scientific issues as part of the science curriculum in which students are to take part and discuss the cause and affects. The teaching of social scientific issues in a science curriculum could help students in developing informal reasoning and argumentation skills. The author explores the relationships between school education, social science issues, and the scientific literacy. The advantages of teaching social scientific issues are many. According to the author, teaching social scientific issues can enhance decision making and critical thinking, promote science communication, induce learning interest about science, and provide cross-disciplinary concepts. Later in the chapter, the author presents findings from previous research on people's attitudes towards social scientific issues. According to the findings, these can be divided into three: factual, value and policy categories. The chapter elaborates on the roles that social science issues play in science education and defines people's attitudes towards such issues based on the above-mentioned categories. With the increase in science-related issues in our daily life, understanding and attitudes toward science are becoming more important. The author discusses these implications and recommends further research in this area.

Welch and Huffman report their findings from research conducted on secondary students' attitudes towards scientists in Chapter 10. The study was conducted using the Test of Science Related Attitudes (TOSRA). The TOSRA is designed to measure seven science-related attitudes. The scales covered in the instrument are Social Implications of Science, Normality of Scientists, Attitudes towards Scientific Inquiry, Adoption of Science Attitudes, Enjoyment of Science Lessons, Leisure Interest in Science and Career Interests in Science. Each scale has ten items, and the total instrument contains 70 items. The questionnaire was administered to a group of students who participated in the school robotic competition and to another group who did not participate in the competition. The results show that the students who participated in the robotic competition display more positive attitudes towards scientists than those who did not. The authors suggest that the robotic competition appears to have helped students see scientists as normal people instead of being classified as smart or "nerds." They concluded that programs which engage students in authentic problems can significantly improve students' attitudes and views of science.

"Attitudes towards Science and Scientific Methodology within a Specific Professional Culture" by Darko Hren from the University of Split investigates the importance of a research methodology course for medical students in Chapter 11. The chapter looks at ongoing research to see the impact of this particular course on student attitude towards science. The author reports the results of a short- and long-term study in his chapter. The short-term study looks at the attitudes of two groups of second-year medical students. The study lasted for four months, and the results showed there were no differences between the two groups. On the other hand, the long-term study showed that attitude of students towards science "increased after they finished the course on scientific methodology, but then decreased in subsequent two years and finally increased again in the final year of their study." He concludes the chapter by pointing out the importance of scientific methodology courses to improve medical students' attitudes towards science.

CONCLUSION

Osborne, Simon, and Collins (2003) urged that measuring students' attitudes towards studying science is an"urgent agenda for research."They reviewed the literature published in the last 20 years about attitudes towards science, its implication and pointed out that such attitudes are influenced by many factors that include gender, teachers, curricular, cultural and other factors. The contributors to this volume have considered challenges and potentials in this area. It is hope that this volume will be a compendium for science educators in using attitudes measurements as a potentially useful and effective educational tool to improve science education.

REFERENCE

Blalock, C., Lichtenstein, M., Owen, S., Pruski, L., Marshall, C., & Toepperwein, M. (2008). In pursuit of validity: A comprehensive review of science attitude instruments 1935–2005. *International Journal of Science Education, 30*(7), 961–977.

Bond, T. G., & Fox, C. M. (2001). *Applying the Rasch model: Fundamental measurement in the human science.* Mahwah, NJ: Lawrence Erlbaum Associates.

Eagly, A. H. & Chaiken, S. (1993). *The psychology of attitudes.* London: Harcourt Brace Jovanovich College.

Gardner, P. L. (1996). The dimensionality of attitude scales: A widely misunderstood idea. *International Journal of Science Education, 18*(8), 913–919.

Hovland, C.I., Janis, I.L., &Kelly, H. (1953). *Communication and persuasion.* New Haven, CT: Yale University Press.

Lederman, N. G. (2007). Nature of science: Past, present and future. In S. K. Abell & N. G. Lederman (Eds.), *Handbook of research on science education* (pp. 831–880). Mahwah, NJ: Lawrence Erlbaum.

Munby, H. (1997). Issues of validity in science attitude measurement. *Journal of Research in Science Teaching, 34*(4), 337–341.

Noll, V.H. (1935). Measuring the scientific attitude. *Journal of Abnormal and Social Psychology, 30*, 1–7.

Osborne, J., Simon, S., & Collins, S. (2003). Attitude towards science: A review of the literature and its implication. *International Journal of Science Education, 25*(9), 1049–1079.

Petty, R.E. & Cacioppo, J.T. (1996). *Attitudes and persuasion: Classical and contemporary approaches.* Boulder, CO: Westview Press.

Reid, N. (2006). Thoughts on attitude measurement. *Research in Science and Technological Education, 24*(1), 3–27.

Schibeci, R. (1984). Attitudes to science: An update. *Studies in Science Education, 11*, 2–59.